UNIPA Springer Series

The **UNIPA Springer Series** publishes single and co-authored thematic collected volumes, monographs, handbooks and advanced textbooks, conference proceedings, professional books, SpringerBriefs, journals on specific issues of particular relevance in six core scientific areas. The issues may be interdisciplinary or within one specific area of interest. Manuscripts are invited for publication in the following fields of study:

1- Clinical Medicine;
2- Biomedical and Life Sciences;
3- Engineering and Physical Sciences;
4- Mathematics, Statistics and Computer Science;
5- Business, Economics and Law;
6- Human, Behavioral and Social Sciences.

Manuscripts submitted to the series are peer reviewed for scientific rigor followed by the usual Springer standards of editing, production, marketing and distribution. The series will allow authors to showcase their research within the context of a dynamic multidisciplinary platform. The series is open to academics from the University of Palermo but also from other universities around the world. Both scientific and teaching contributions are welcome in this series. The editorial products are addressed to researchers and students and will be published in the English language.

More information about this series at http://www.springer.com/series/13175

Matteo Campanella · David Jou ·
Maria Stella Mongiovì

Interpretative Aspects of Quantum Mechanics

Matteo Campanella's Mathematical Studies

UNIVERSITÀ
DEGLI STUDI
DI PALERMO

Springer

Foreword

Matteo Campanella graduated cum laude in Electronic Engineering at the University of Palermo in May 1972. He has been an Assistant Professor at the same university since 1973. He became an Associate Professor on Fundamentals of Electrical Engineering in 1983 and on Telecommunications in 1992. Since 2002, he has been a Full Professor on Telecommunications at the University of Palermo.

Matteo Campanella has been always a very eclectic researcher, as demonstrated not only by the heterogeneity of his research interests, but also by the variety of subjects that he taught during his career. He started with a class about Synthesis of Passive and Active Circuits in 1973, then he taught Applied Electronics from 1974 to 1978 and Circuit Theory from 1974 to 1991. Although since 1983 he was an Associate Professor on Fundamentals of Electrical Engineering, with the official course on Circuit Theory, in 1980, 1987, and 1988 he taught the course of Electromagnetic Fields and Circuits, and in 1985–1986 he taught again the course of Applied Electronics. In 1991, he taught for the first time the course of Numerical Transmission Systems that he kept during the rest of his career. Since 1997, he also taught the course of Digital Processing for the Master Program of Telecommunication Engineering and Computer Engineering.

He participated in the Ph.D. course on Communications at the University of Palermo and he was the advisor of several master thesis and Ph.D. thesis dealing with distributed circuits and numerical transmission systems. He devoted his studies to the theory of microwave circuits, optimization techniques for continuous phase modulated systems, evaluation of limit performance of constant envelop channels as well as codes and lattices, and convolutional codes over groups and turbo codes. He has been reviewer for several important journals, among which some of the journals are IEEE Transactions on Communications, IEEE Transactions on Information Theory, and IEEE Communications Magazine. He also worked on several national research programs, funded by the Italian Minister of University and Research (MIUR).

Matteo Campanella was a very intellectually gifted researcher, but also a very charismatic lecturer. He delighted colleagues and students with his wide knowledge and vibrant passion for ideas as well as the strength of his coherence. Although he was a very reserved and modest person, his mild temperament and strong personality, as well as his way of facing complex problems as a sequence of obvious intuitive steps, provided that good analysis instruments were defined, significantly affected hundreds of students and the whole group of Telecommunications at the University of Palermo.

Palermo, Italy
January 2018

Pierluigi Gallo
Giovanni Garbo
Giovanni Mamola
Stefano Mangione
Ilenia Tinnirello

Preface

Professor Matteo Campanella was born in Palermo on September 27, 1947, and died prematurely on June 18, 2016. He was a Professor at the DEIM Department (Dipartimento di Energia, ingegneria dell'Informazione e modelli Matematici), Palermo University. This monograph presents a selection of his writings on the foundations on quantum mechanics and on a possible derivation of the Born rule. The notes are presented just as he wrote them. Professor Matteo Campanella planned to explore some of these ideas in more depth, but could not complete his work, because of his premature death.

I met my husband Matteo in 1982, and was immediately fascinated by his personality. His cultural interests ranged from the sciences to music, history, and philosophy. He was just as passionate about the sciences as he was about political and social issues. He loved physics and mathematics. He had been studying the foundations of quantum mechanics and, in particular, its interpretation, for years. Sometimes he spoke with me about these topics, and I always tried to encourage him to publish his work. In 2011, he told me about an interesting new result. This time, I really tried to persuade him to publish his results, but he wasn't fully satisfied with what he'd written, and intended to modify some parts of it. Five years passed: I often asked him how his work was going, and he always told me that it wasn't entirely clear. Over these 5 years, he continued to study the problem. The results of his studies can now be found in over 100 files from his computer, and in a wealth of written notes. I decided to publish a part of his writings on this subject, hoping that someone will someday continue his work.

Palermo, Italy

Maria Stella Mongiovì

Introduction

This monograph contains some posthumous mathematical writings of Prof. Matteo Campanella on some interpretative aspects of quantum mechanics. The general goal of this work is to arrive to Born's rule, one of the key principles of the probabilistic interpretation of quantum mechanics, *in a way independent of any a priori probabilistic interpretation*. This topic is indeed a very active line of research nowadays, because of its fundamental interest. Here we try to outline and summarize the main lines of this work. To simplify the reading of the text, we have split each chapter in a few sections; this division was not introduced by Campanella himself, and it is only aimed to provide an indicative guide to the main lines of the content.

General Context

Born's rule was proposed by Max Born in 1926, in an attempt to interpret the physical meaning of the wave function introduced by Schrödinger, by relating it to the probability of obtaining a particular value in the set of the allowed values of the physical quantities characterizing the system, when making a measurement on the system. It thus plays an essential role in the connection between quantum theory and experiments and it has a deep conceptual interest because it was one of the ways in which indeterminism entered in fundamental physics, in parallel with Heisenberg's relations. In the standard formalism of quantum mechanics, it is considered as a basic postulate. However, due to its deep implications and its operational role in measurement theory, its physical derivation and its philosophical interpretation have always deserved much interest [18, 20, 24, 192, 188, 191, 206, 236, 198]. Recall, for instance, that the same idea of probabilistic interpretation of the wave function aroused much discussions between Einstein, for whom this probabilistic character was an indication of some missing information about the actual physical world, and Bohr and many other researchers, for whom this probabilistic character was inherent of deep physical reality, and quantum mechanics was a complete theory.

Many attempts toward a consistent derivation of Born's rule have been made along several different lines, for instance, in the many-universes interpretation (since 1957) [131, 197, 134], in the de Broglie-Bohm interpretation (since 1920) [6, 187, 5, 139, 146, 145, 244, 127, 140], or from Gleason's theorem (since 1957) [9, 207, 224]. At the turning of the twenty-first century, other kinds of proposals arised, stimulated by a stronger emphasis on the idea of decoherence and of entanglement and on the fast expansion of research on quantum information. Let us mention a few approaches: Deutsch proposed to derive Born's probabilistic rule from the non-probabilistic axioms of quantum mechanics combined with classical decision theory [39, 61, 11, 40, 41]; Zurek proposed an environment-related derivation based on the invariance (relative to a given environment $\mathscr{E}$, and therefore called "envariance" in short) of entangled states under a swap of the outcomes of the system $\mathscr{S}$ without changing the outcomes of the environment $\mathscr{E}$ [248, 249, 251, 53, 250, 87, 23, 27]. Zurek derivation has been examined by several authors [212, 221, 188, 211, 224, 10]. Relational quantum mechanics (since 1994) [180, 183, 182, 181], or categorical quantum mechanics (since 2005) [31, 29, 28, 30], especially stressing the role of relations or of processes or Bayesian quantum mechanics [194], intuitionistic quantum mechanics [232], Ithaca approximation [186, 104] are other contemporary approaches of giving new interpretation to several basic aspects of quantum foundations and quantum logic.

The Present Work

Campanella's work participates to this general trend of research and it aims to giving a more rigorous mathematically based form to proposals in the line set forth by Zurek since 1992.

Quantum mechanics in general, and the mentioned derivations in particular, make several important mathematical assumptions about the definition of state of the system, the invariant transformations related to it, the relation between the state and probabilities, and so on. The aim of Campanella's work is to start from the uncontroversial interpretative assumptions of quantum mechanics and to use a suitable mathematical formalism, to derive the other most relevant interpretative items of quantum mechanics, especially Born's rule.

In Chap. 1, the general assumptions and aims of the work are presented. The mathematical assumptions underlying the interpretative items of quantum mechanics are: 0) the properties of a quantum system H can be described in terms of a Hilbert space H associated with it; 1) if the system W is a universe, a state of it may be identified as a ray of W, that is, an element of a projective space $\mathbb{P}(W)$; 2) if a system L is composed of two subsystems H and K its Hilbert space is a tensor product $H \otimes K$; 3) the only allowed transformations for the states of a universe W

are induced by unitary transformations of its Hilbert space W. These are well-accepted features in quantum mechanics. At this point, an essential part of Campanella's program is the use of a formulation mathematically equivalent to transformations but involving a single state of an extended composite system which, seen from different partial points of view, appears as the transformation, the state "before" the transformation and the state "after" the transformation. In doing so, it is intended to avoid two drawbacks of a transformation involving a) the totality of states and b) the idea of an "evolution" from an old state to a new one.

Chapters 2–3 are the central ones. In Chap. 2, *the state of a subsystem of a quantum system is characterized independently of any probabilistic interpretation.* In contrast to classical mechanics, in quantum mechanics the nature of a state of a subsystem of a given composite system is deeply different as that of the whole system, because of the presence of entanglement for the states. In the conventional setting of quantum mechanics, the state of the subsystem is described by a partial density operator, whose definition is justified by the probabilistic interpretation and the use of Born's rule. In Chap. 2, instead, it is shown that *one way of characterizing the state of a subsystem is through a partial density operator independently of any a priori probabilistic interpretation.* The latter will come out as a consequence of the formalism and of a few assumptions, connected with the notion of a state. Suppose that a composite system $\mathscr{C}$ consists of two parts $\mathscr{S}$ (the "system") and $\mathscr{E}$ (the "environment") with respective Hilbert spaces H and K. A key step for the introduction of a natural definition of a state of $\mathscr{S}$ is the *formalization of the "envariance" property introduced by Zurek* (namely, an "environment-assisted" invariance exhibited by entangled systems). From this assertion, and describing the states of the composite system ("the universe") as pure states, it is concluded that the state of $\mathscr{S}$ associated with a pure state $|\phi\rangle$ of the composite system may be defined as the orbit of $|\phi\rangle$ under the action of the group $\mathbf{I} \otimes \mathscr{U}(K)$, being I the identity operator and $\mathscr{U}(K)$ the group of the unitary transformations in K.

According to this, the state of $\mathscr{S}$ is characterized as a subset of the space of the composite system. To arrive at a definition involving only points of the Hilbert space of $\mathscr{S}$, any set of mathematical objects in bijection with the orbits of $\mathbf{I} \otimes \mathscr{U}(K)$ can be as well used to define the state. Thus, the author arrives at a characterization of a state of $\mathscr{S}$ as a *density operator arising as an orbit invariant without any a priori probabilistic interpretation.*

In Chap. 3, the step from the above non-probabilistic definition of the state of $\mathscr{S}$ to a probabilistic definition is given. Two related, but distinct items about this discussion are considered. The first consists in the *different ways of representing the state of $\mathscr{S}$ depending on whether it is considered isolated or not.* Classically, the state of a composite system is described as a point in its phase space, expressed as the Cartesian product of the phase spaces of the component subsystems. If also in the classical case the state of $\mathscr{S}$ is defined a la Zurek, with the role of $\mathscr{U}^{Z}(K)$ substituted by the role of the group of the canonical transformations of the phase space of $\mathscr{E}$, the orbits of the latter group can be parameterized by the points of the phase space of $\mathscr{S}$ both in the absence and in the presence of interaction. So in the classical case the conventional definition of states of the subsystem is compatible with Zurek's point of view. In the latter case, the projections of the initial and final points of the composite system are points of the phase space of $\mathcal{S}$ even if an interaction takes place. Since the final point of $\mathcal{S}$ depends both on the initial states of $\mathcal{S}$ and of $\mathcal{E}$, if the state of the environment $\mathcal{E}$ is not completely known statistical methods must be introduced, but the description of the state of $\mathcal{S}$ by means of a point in its phase space is not forbidden, at least in principle. The situation is quite different in the quantum case. Even if the initial state is described by a point of $\mathbb{P}(H)$, the final state of $\mathcal{S}$ (i.e., the "projection" of the final state of the composite system) is a density operator with rank > 1 and not just a one-dimensional projector if an interaction with the environment has taken place.

The second question concerns the *different ways of describing the evolution of the system in the absence or in the presence of interaction*. Both in classical and quantum cases the interaction precludes the possibility of establishing a functional relationship between the initial and the final states of $\mathcal{S}$: the only thing that can be said, in general, is that both are "projections" of the initial and final states of the composite system. But, in contrast to the classical case, in the quantum case a fundamental difference exists in the mathematical description of the states of $\mathcal{S}$ depending on whether the system is considered as isolated or as a subsystem of $\mathcal{C}$. In the first case, the states are points of $\mathbb{P}(H)$, while in the second one they are represented by density operators. *In the most popular axiomatics of quantum mechanics, the need of this difference is a consequence of the a priori introduction of a probabilistic interpretation* of the theory and its quantitative expression through Born's rule. Instead, *here this difference is a direct consequence of two assumptions*: that a composite system is described by a tensor product and that the correct notion of state of a subsystem is the one implied by Zurek's considerations. *Neither of these assumptions involves probability*; specifically, *the density operator arises merely as an orbit invariant.*

Suppose we may consider the system $\mathcal{S}$ isolated from the environment $\mathcal{E}$ before some instant t_0, but that the interaction is switched on, until t_1, so that, for $t > t_1 \mathcal{S}$ and $\mathcal{E}$ can be again modeled as mutually isolated. Before t_0, $\mathcal{S}$ and $\mathcal{E}$ evolve independently; at $t<t_0$, $\mathcal{S}$ is in a well-definite state $\varsigma \in \mathbb{P}(H)$ and $\mathcal{E}$ in a state $\tau \in \mathbb{P}(K)$. At t_0 the interaction is switched on. Therefore, the state of the composite system $\mathcal{C}$ after t_0 becomes entangled. In particular, let $|\psi\rangle$ denote a normalized representative of the state at t_1. At this instant the system must be considered as interacting with the environment as the "past" is concerned, and as non-interacting regards to the "future." If "non-interacting" were the same thing as "isolated," we could think that both descriptions of the state must hold: as an interacting system, it ought to be described by $\rho_S = tr_E|\psi\rangle\langle\psi|$, (which means a partial specification of the state of $\mathcal{C}$) and, as a non-interacting system, by a (perhaps partial) specification of some point of ξ of $\mathbb{P}(H)$.

A careful analysis shows that, when there is in ρ_S some kind of degeneracy, the notions of "non-interacting" and "isolated" are not equivalent. This means that not for all the density operators the description in terms of states of an isolated system is possible. More precisely, it is shown that if a system $\mathcal{H}$ can be considered isolated there is at most a single multidimensional eigenspace of ρ_H and that, if it exists, it is the null eigenspace. The spectral expansion of ρ_H is $\rho_H = \sum wP(w)$ where all the projectors $P(w)$ are of rank one, and the possible states of $\mathscr{H}$ in ρ_H as an isolated system are the states associated with these projectors.

In the next step, it is seen that *we are forced to give to a density operator whose non-null eigenspaces are all one-dimensional, called generic density operator, a probabilistic interpretation*: it is shown that it is possible to fix uniquely a probability law on the set Ξ of eigenstates of ρ_S, so that the state becomes a well-defined random variable.

This result is obtained in the last chapter (Chap. 4). The lines of thought to obtain this are the following. In Sect. 4.1, the structure of the set D_Ξ of generic density operators is analyzed. Barycentric coordinates with respect to an orthogonal set of states are introduced, bringing a simplex structure together with the standard topology. After a detailed mathematical analysis of the set D_Ξ, in Sect. 4.2, the problem of the probability of the non-null eigenstates of a generic density operator is studied and the universality of the probability distribution is shown. Sections 4.3 and 4.4 are devoted to the determination of a functional equation for the probability distribution function and of its solution. One sees that the only solution for the obtained functional equation is the identity. One concludes that if two operators have the same spectrum, they coincide. This final result assures that, without any additional axiom, *the probability of each state is shown to be equal to the corresponding eigenvalue of the density operator.* Thus, Born's rule is recovered as a consequence rather than introduced as an axiom.

The appendices discuss many mathematical details which would have broken the continuity of the discussion of the former chapters if they had been included at the corresponding points. For instance, they deal with barycentric coordinates, mapping between Hilbert spaces, tensor products between linear spaces, orbits of vectors of a linear space, under the action of its structure group and the class of Hilbert space as a category. In particular, in Appendix D, the proof of the Theorem 2.4, that is, an important stem in Campanella's discussion, can be found.

In the writings of Prof. Campanella, there was not an explicit bibliography at well-defined points. Instead, the bibliography he used for his work was collected in several files. To be as faithful as possible to his work we present the bibliography as it was structured in his several files. For the sake of completeness regarding the general context, we have added in an independent list other references related to this topic.

David Jou
Maria Stella Mongiovì

Contents

1 Fundamental Assumptions 1
1.1 Basic Assumptions of Quantum Mechanics 1
1.2 Quantum States as Rays and Related Mathematical Concepts 2

2 The State of a Quantum System as a Subsystem of a Composite System 9
2.1 Non-probabilistic Characterization of the State of a Subsystem . . . 9
2.2 Mathematical Characterization of Zurek's Envariance 10
2.3 Characterization of a State Through Its Properties 17

3 Relation Between the State of a System as Isolated and as Open . . . 21
3.1 Representation of Isolated and Non-isolated Systems 22
3.2 Conditions for Density Operators. 24
3.3 Physical Interpretation of Generic Density Operators 30

4 The Probability Law for Generic Density Operators 33
4.1 Mathematical Structure of a "Generic" Density Operator. 33
4.2 Universality of the Probability Distribution Function of a Generic Density Operator. 39
4.3 A Functional Equation for g_2 43
4.4 Solution of the Functional Equation. 47
4.5 The Function g for Arbitrary Rank 51

Appendix A: Categories 57

Appendix B: Barycentric Coordinates. 77

Appendix C: Linear Spaces 85

Appendix D: Mathematical Frameworks. 103

References 131

Chapter 1
Fundamental Assumptions

1.1 Basic Assumptions of Quantum Mechanics

The approach to our discussion will be roughly the following. We will state without further analysis the interpretative assumptions of Quantum Mechanics which are at present substantially uncontroversial. On the contrary, we will try to derive from them and from the mathematical formalism the other most relevant interpretative items (including Born's rule).

The first assumption we make without discussion is that, *if a system W is considered as a universe, a state of it can be specified by a ray of W, that is an element of the associated projective space $\mathbb{P}(W)$.*

An equivalent formulation is to specify the state through a projector P of rank one operating on W. We will denote ψ an element of $\mathbb{P}(W)$ and $|\psi\rangle$ any normalized ket representing ψ (defined up to an arbitrary phase factor). Consequently, the corresponding projector is $P = |\psi\rangle\langle\psi|$.

The second assumption is that, *if a system L is composed with the two subsystems H and K, its Hilbert space is a tensor product $H \otimes K$ (as a Hilbert space).*

We observe that a tensor product of two spaces is, strictly speaking, not a space, but a bilinear mapping whose domain is the Cartesian product $H \times K$. Hence, we ought to say that the Hilbert space of L is the image of a tensor product. Furthermore, tensor products are defined up to isomorphisms. The choice of one or another realization of the tensor product is not essential, because we can translate in a one-to-one way any statement expressed in a realization into a statement expressed in any other.

However, a natural and, for many instances, convenient realization of $H \otimes K$ in the case of (finite-dimensional) Hilbert spaces is the linear space $Hom\,(H^*, K)$ of the linear mappings from the dual H^* of the space H to the space K. The scalar product $\langle\varphi, \psi\rangle$ is defined as $tr\,(\varphi^* \circ \psi)$, where φ^* denotes the adjoint of φ [see Appendices C and D]. A mapping of rank one of $Hom\,(H^*, K)$ is expressed as $|k\rangle_{K\,H^*}\langle h|$. We have

Note: The content of this chapter can be found in Campanella's file *Zurekabs (16-12-2011)*.

M. Campanella et al., *Interpretative Aspects of Quantum Mechanics*, UNIPA Springer Series,
https://doi.org/10.1007/978-3-030-44207-1_1

$$(|k\rangle_{K\,H^*}\langle h|)\,(|x\rangle_{H^*}) = |k\rangle_{K\,H^*}\langle h|x\rangle_{H^*} = {}_H\langle x|h\rangle_H|k\rangle_K.$$

Simplifying the notation, an element of rank one will be expressed as $|h\rangle\,|k\rangle$, with the rule of calculation

$$(|k\rangle\ _{H^*}\langle h|)\,(|x\rangle_{H^*}) = \langle x|h\rangle\ |k\rangle\,.$$

We put $|h\rangle \otimes |k\rangle = |h\rangle\,|k\rangle$. Hence, in this realization, the separable elements of $H \otimes K$ are the linear mappings of rank one. Unless differently specified, the above realization of a tensor product will be understood.

The third uncontroversial notion is that *the only allowed transformations for the states of a universe W are induced by unitary transformations of its Hilbert space. This means that an allowed transformation of the elements of $\mathbb{P}(W)$ is induced by an element of the group $\mathfrak{U}(W)$ of the unitary transformations of W.*

1.2 Quantum States as Rays and Related Mathematical Concepts

The notion of a transformation implies a mapping, which associates to any state "before" a corresponding state "after." This point of view has two drawbacks: it involves the totality of states and it involves the idea of an "evolution" from an old state to a new one. We would like to arrive at a mathematically equivalent formulation which involves a single state (of an extended composite system) which, seen by different partial points of views, appears as the transformation, the state "before" and the state "after." A more precise sense of what we mean will come out from the following analysis.

The key observation which allows to carry on this program is the bijection between linear mappings and elements of tensor products. To the space of linear mappings $Hom\,(X, Y)$, we associate the tensor product $X^* \otimes Y$. Using the canonical isomorphism between $X^* \otimes Y$ and $L = Y \otimes X^*$, we associate a ket of the latter space to a specific linear mapping, which can be considered as a representative of a state of a composite system L. If Z is another space, $\psi \in Hom\,(X, Y)$ and $\varphi \in Hom\,(Y, Z)$, the composition $\varphi \circ \psi$ is defined and belongs to $Hom\,(X, Z)$. Introducing the tensor products $M = Z \otimes Y^*$ and $N = Z \otimes X^*$, we associate to ψ and φ the corresponding kets $|\psi\rangle \in L$ and $|\varphi\rangle \in M$. The ket $|\varphi \circ \psi\rangle$ depends linearly on $\varphi \circ \psi$ as well as $|\psi\rangle$ and do on ψ and φ. Hence, the mapping which associates to the pair $(|\varphi\rangle\,, |\psi\rangle)$ the ket $|\varphi \circ \psi\rangle$ is bilinear. Consequently, there is a unique linear mapping

$$M \otimes L \xrightarrow{\kappa_Y} N \qquad \text{such that} \qquad |\varphi \circ \psi\rangle = \kappa_Y\,(|\varphi\rangle \otimes |\psi\rangle)\,. \tag{1.2.1}$$

There is a canonical isomorphism between the spaces $M \otimes L$ and $Z \otimes Y^* \otimes Y \otimes X^*$. With an abuse of notation we will understand $|\varphi\rangle \otimes |\psi\rangle$ as an element of the latter space, and κ_Y with the same space as its domain. The application of κ_Y will be called *contraction* in Y.

Conversely, given a ket $|\chi\rangle \in Z \otimes Y^* \otimes Y \otimes X^*$ admitting a factorization

$$|\chi\rangle = |\varphi\rangle \otimes |\psi\rangle, \qquad \text{with} \qquad |\varphi\rangle \in Z \otimes Y^* \ \text{ and } \ |\psi\rangle \in Y \otimes X^*,$$

the factors $|\varphi\rangle$ and $|\psi\rangle$ are determined up to scalar factors λ and μ such that $\lambda\mu = 1$. From the object $|\chi\rangle$, we can thus recover the ket $|\varphi \circ \psi\rangle$, but not the factors of its decomposition because of the above indeterminacy. The latter is however removed if we pass to the corresponding rays.

A ray of morphisms belonging to $Hom\,(X, Y)$ can be represented as a ray of $Y \otimes X^*$ and a ray of $Hom\,(V, Z)$ can be represented as a ray of $Z \otimes V^*$. A pair of such rays can be represented in a one-to-one way as a ray of $Z \otimes V^* \otimes Y \otimes X^*$. If $V = Y$ the composition of morphisms is possible and the corresponding ray is obtained contracting on Y.

If a ray of $Z \otimes V^* \otimes Y \otimes X^*$ is the corresponding of a pair of rays of $Z \otimes V^*$ and $Y \otimes X^*$, we would like to recover the original rays through contractions. This is not possible in a direct way, but we can proceed as follows. To a ray represented by $|x\rangle \in X$ we can associate the ray generated by the element $|x\rangle\,\langle x| \in Hom\,(X, X)$, which corresponds to the element $|x\rangle \otimes \langle x| \in X \otimes X^*$. The original ray can be uniquely recovered by the ray generated by the latter element.

For our purposes, it is more useful to work with the ray generated by $|x\rangle \otimes \langle x|$ rather than with the original ray. In our case, if a ket $|\psi\rangle$ represents a ray of $Y \otimes X^*$, we will use instead a ray generated by $|\psi\rangle \otimes \langle\psi|$ which belongs to $Y \otimes X^* \otimes X \otimes Y^*$. Similarly, if a ray of $Z \otimes V^*$ is represented by $|\varphi\rangle$, we will use the ray generated by $|\varphi\rangle \otimes \langle\varphi|$ which belongs to $Z \otimes V^* \otimes V \otimes Z^*$. The pair $(|\varphi\rangle\,, |\psi\rangle)$ is represented by an element of $Z \otimes V^* \otimes V \otimes Z^* \otimes Y \otimes X^* \otimes X \otimes Y^*$. The ray generated by $|\varphi\rangle \otimes \langle\varphi|$ can be recovered contracting on X and then on Y, while the ray generated by $|\psi\rangle \otimes \langle\psi|$ is obtained contracting on V and then on Z. Finally, if $V = Y$, the ray generated by $|\varphi \circ \psi\rangle \otimes \langle\varphi \circ \psi|$ is obtained by contracting on V^* and Y and on V and Y^*.

In order to be able to deal in a simple way with situations of arbitrary complexity, we need to build a general formalism which is similar to tensor algebra. The main difference is that, while in the latter we deal with iterated tensor products obtained starting from a fixed space and its dual, here we build tensor products of arbitrary (finite-dimensional) spaces.

We remember that to each ring we can associate a module in the following way.

Let $(R, +, \cdot)$ be a ring. Let $\{\frown\}$ be a singleton disjoint from R and let $\widehat{R} = R \times \{\frown\}$. We will denote $\widehat{r}$ the pair $(r, \frown)$. We introduce in $\widehat{R}$ a structure of additive group putting $\widehat{r} + \widehat{s} = \widehat{r + s}$. We further define $r\widehat{s} = \widehat{rs}$. With the above definitions, $\widehat{R}$ is equipped with a structure of R-module. In particular, if R is a field, $\widehat{R}$ is a linear space over R.

If we take $R = \mathbb{C}$, $\widehat{\mathbb{C}}$ is a complex linear space. Usually, an improper terminology is used, saying that $\mathbb{C}$ can be regarded a complex linear space over itself. The precise meaning of this terminology is the above construction.

For any linear space X over $\mathbb{C}$, its dual is defined as the space of linear mappings from X to $\mathbb{C}$. An element of this space is thus a mapping

$$X \xrightarrow{\varphi} \mathbb{C} \quad \text{such that} \quad \varphi(\lambda x + \mu y) = \lambda\varphi(x) + \mu\varphi(y). \tag{1.2.2}$$

If we define the mapping $\widehat{\varphi}$ through the position $\widehat{\varphi}(x) = (\varphi(x), \widehat{\ })$, we obtain an element of $Hom(X, \widehat{\mathbb{C}})$. We redefine the dual space as the linear space $X^* = Hom(X, \widehat{\mathbb{C}})$ and neglect the hat in $\widehat{\varphi}$.

There is a *natural linear mapping* $X \xrightarrow{\iota_X} X^{**}$ defined as follows. The expression $x^*(x)$ defines a mapping $X^* \times X \to \widehat{\mathbb{C}}$ which is linear in both arguments. In particular, for each fixed x we obtain a linear mapping from X^* to $\widehat{\mathbb{C}}$, that is, an element $\iota(x)$ of X^{**}.

We recall that a *basis* of a linear space X is a set mapping $B \xrightarrow{\beta} X$ such that (β, X) is a free linear space, and that X is finite-dimensional if there is some basis such that $|B|$ is finite. It is well known that a linear space always has a basis and that all bases have the same cardinality. In the finite-dimensional case, the natural number $|B|$ is the dimension of the space. We emphasize that a basis is a mapping and not a set. The image of a basis will be called a basis set. Given a basis $B \xrightarrow{\beta} X$, we can define the dual basis $B \xrightarrow{\beta^*} X^*$ as follows. The position

$$\beta^*(b)\left(\beta(b')\right) = \delta_{bb'}\widehat{1}$$

uniquely defines an element $\beta^*(b)$ of X^* as

$$\beta^*(b)(x) = \beta^*(b)\left(\sum x(b')\,\beta(b')\right) = x(b)\,\widehat{1}.$$

The mapping ι_X is an isomorphism in the case of finite-dimensional spaces.

If $\varphi \in Hom(X, Y)$, we can define the *transpose* $\varphi^t \in Hom(Y^*, X^*)$ as follows. For each $y^* \in Y^*$, the mapping $y^* \circ \varphi$ is an element of $Hom(X, \widehat{\mathbb{C}})$, that is, of X^*. As y^* varies in Y^*, we get a linear mapping, that is, an element φ^t of $Hom(Y^*, X^*)$. The mapping $(\ )^t$ is a linear mapping from $Hom(X, Y)$ to $Hom(Y^*, X^*)$. The mapping φ^{tt} is an element of $Hom(X^{**}, Y^{**})$. Consequently, $\iota_Y^{-1} \circ \varphi^{tt} \circ \iota_X$ is defined. It can be easily shown that $\iota_Y^{-1} \circ \varphi^{tt} \circ \iota_X = \varphi$.

Let us apply the above notions to the space $\widehat{\mathbb{C}}$. The set injection

$$\{1\} \to \widehat{\mathbb{C}} \ : \ 1 \mapsto \widehat{1}$$

equips $\widehat{\mathbb{C}}$ with a structure of free space over $\{1\}$, so that this mapping is a basis of $\widehat{\mathbb{C}}$. Let f be an element of $\widehat{\mathbb{C}}^*$. If $\widehat{x} \in \widehat{\mathbb{C}}$, $\widehat{x} = x\widehat{1}$, so that

$$f(\widehat{x}) = xf\left(\hat{1}\right) = x\lambda\widehat{1},$$

where $\lambda\widehat{1} \triangleq f(\widehat{1})$. Hence $f(\widehat{x}) = \lambda\widehat{x}$. Conversely, for each $\lambda \in \widehat{\mathbb{C}}$ the above position defines an element of $\widehat{\mathbb{C}}^*$.

Let $\mathbb{C} \xrightarrow{\tilde{\vartheta}} \widehat{\mathbb{C}}^{\mathbb{C}}$ be the mapping defined by

$$\tilde{\vartheta}(\lambda)(\widehat{x}) = \lambda\widehat{x}.$$

Then $\widehat{\mathbb{C}}^* = \mathrm{Im}\,\tilde{\vartheta}$ and the restriction ϑ of $\tilde{\vartheta}$ to its image is a bijection. We have

$$x^*(x) = \vartheta^{-1}\left(x^*\right)x.$$

Remembering that $\iota_X(x)(x^*) = x^*(x)$, we get $\iota_X(x)(x^*) = \vartheta^{-1}(x^*)\,x$ so that

$$\iota_X = \cdot \circ \vartheta^{-1},$$

where "$\cdot$" denotes the action of $\mathbb{C}$ on $\widehat{\mathbb{C}}$ defined by the scalar multiplication. Each element of the image of ι_X is a mapping $\mathbb{C} \xrightarrow{\gamma_x} \widehat{\mathbb{C}}$ defined by $\gamma_x(\lambda) = \lambda x$. For each x, we get an element of $\widehat{\mathbb{C}}^{\mathbb{C}}$. Hence, we have a mapping $\widehat{\mathbb{C}} \xrightarrow{\tilde{\gamma}} \widehat{\mathbb{C}}^{\mathbb{C}}$. Hence

$$\widehat{\mathbb{C}}^{**} = \mathrm{Im}\,\tilde{\gamma} \subseteq \widehat{\mathbb{C}}^{\mathbb{C}}.$$

Hence, while the elements of $\widehat{\mathbb{C}}^*$ are all the mappings from $\widehat{\mathbb{C}}$ to itself corresponding to a scalar multiplication, the elements of $\mathbb{C}^{**}$ are all the mappings from $\mathbb{C}$ to $\widehat{\mathbb{C}}$ corresponding to a multiplication by an element of $\widehat{\mathbb{C}}$.

We can consider the mapping $\mathbb{C} \times \widehat{\mathbb{C}} \to \widehat{\mathbb{C}}$. For each fixed $\lambda \in \mathbb{C}$, we get an element of $\mathbb{C}^*$ defined by the rule $x \mapsto \lambda \cdot x$, while for each fixed x we get an element $\iota_X(x)$ of $\mathbb{C}^{**}$ defined by the rule $\lambda \mapsto \lambda \cdot x$. The dual basis of $1 \mapsto \widehat{1}$ is $1 \mapsto \widehat{1}^*$ where $\widehat{1}^*(\widehat{1}) = \widehat{1}$, so that $\widehat{1}^*(\widehat{x}) = x\widehat{1} = \widehat{x}$. The position

$$1 \mapsto \widehat{1} \mapsto \iota(\widehat{1}) \triangleq \widehat{1}^{**}$$

equips $\mathbb{C}^{**}$ with the structure of a free space over $\{1\}$.

Defining $\langle\widehat{x}|\widehat{y}\rangle = x^*y$, all the properties of a scalar product are satisfied, so that $\widehat{\mathbb{C}}$ becomes a Hilbert space. Hence, any element $f \in \widehat{\mathbb{C}}^*$ can be uniquely represented as $f(\widehat{y}) = \langle\widehat{x}|\widehat{y}\rangle = x^*y$. In the bra-ket notation, an element of $\widehat{\mathbb{C}}$ is denoted as a ket, and an element of $\widehat{\mathbb{C}}^*$ as a bra. This notation allows the elimination of hats. Indeed, we can put $\widehat{y} = |y\rangle$. In this notation, the symbol inside the ket denotes a complex number. We can say that an element of $\widehat{\mathbb{C}}$ is a complex number "dressed with a ket-suit." The sum of two ket-dressed complex numbers is the sum of the undressed complex numbers, dressed with a ket-suit, that is, $|x\rangle + |y\rangle = |x + y\rangle$. The product $\lambda\,|x\rangle$ is the ket $|\lambda x\rangle$. The bra corresponding to $|x\rangle$, and consistently denoted with $\langle x|$ (the same complex number, dressed with a bra-suit), is the linear functional defined by the position $\langle x|\,(|y\rangle) = \langle x|y\rangle = x^*y$. The ket $|1\rangle$ is a basis of $\widehat{\mathbb{C}}$ because $|x\rangle = |x1\rangle = x\,|1\rangle$. Hence, there is in $\widehat{\mathbb{C}}$ a canonical basis. We can also say that, defining the set injection

$$\{1\} \xrightarrow{\iota} \widehat{\mathbb{C}} : 1 \mapsto |1\rangle\,,$$

$\mathbb{C}$ canonically defines a structure $(\iota, \widehat{\mathbb{C}})$ of free linear space over the set $\{1\}$. Similarly, the set injection

$$\{1\} \xrightarrow{\iota^*} \widehat{\mathbb{C}}^* : 1 \mapsto \langle 1|$$

defines a structure $(\iota^*, \widehat{\mathbb{C}}^*)$ of free linear space over the set $\{1\}$.

The above structure allows the definition of a *canonical isomorphism* between any (finite-dimensional) Hilbert space H and the Hilbert space $Hom(\widehat{\mathbb{C}}, H)$. Indeed, for any $\varphi \in Hom(\widehat{\mathbb{C}}, H)$, we put

$$\upsilon(\varphi) = \varphi(|1\rangle)\,.$$

This mapping is linear. The condition $\upsilon(\varphi) = 0$ entails $\varphi(|c\rangle) = c\varphi(|1\rangle) = 0$. Furthermore, if, for any $|h\rangle \in H$, we put $\varphi(|c\rangle) = c\,|h\rangle$, we get $\upsilon(\varphi) = |h\rangle$. This shows that υ is an isomorphism. Its inverse υ^{-1} is expressed as $\upsilon^{-1}(|h\rangle) = |h\rangle\langle 1|$. Its adjoint is $|1\rangle\langle h|$, so that its squared norm is $tr\,(|1\rangle\langle h|h\rangle\langle 1|) = \langle h|h\rangle$. We conclude that υ is an isometry. Using this isometry, we can replace each Hilbert space H with the space $Hom\left(\widehat{\mathbb{C}}, H\right)$ and its dual H^* with $Hom\left(H, \widehat{\mathbb{C}}\right)$. Hence, we replace $|h\rangle$ with $|h\rangle\langle 1|$ and $\langle h|$ with $|1\rangle\langle h|$.

In general, the space $Hom\,(H, K)$ can be considered a realization of the tensor product $K \otimes H^*$. In this realization, a rank one element $|k\rangle\langle h|$ is regarded as the separable element $|k\rangle \otimes \langle h|$. The first notation for a rank one element will be called the "dyadic notation," while the second one will be called the "tensor notation." We observe that the first notation may give rise to confusion in some circumstances. Namely, suppose that $H = K$. A rank one element can be written in the form $\left|h'\right\rangle\langle h|$. Suppose now that H is the adjoint of some space L, that is $H = L^*$. Then $\left|h'\right\rangle_H = {}_L\left\langle l'\right|$ and ${}_H\langle h| = |l\rangle_L$. The element is then written as ${}_L\left\langle l'\right|\,|l\rangle_L$. This is a correct notation, but we must maintain the double bar, otherwise the meaning is different: it represents the scalar product. In similar circumstances, the bra-ket notation becomes cumbersome.

For any space X, there is a canonical isomorphism with $X \otimes \widehat{\mathbb{C}}^*$. Furthermore, there is a canonical isomorphism between X and $(X^*)^*$. But the latter is $Hom(X^*, \widehat{\mathbb{C}})$, that is, $\widehat{\mathbb{C}} \otimes X^{**}$, which is canonically isomorphic to $\widehat{\mathbb{C}} \otimes X$. Hence, there is a canonical isomorphism between $X \otimes \widehat{\mathbb{C}}^*$ and $\widehat{\mathbb{C}} \otimes X$. In this isomorphism, the elements $|x\rangle \otimes \langle 1|$ and $|1\rangle \otimes |x\rangle$ correspond to each other.

An element of a space X will be regarded as an element of $X \otimes \widehat{\mathbb{C}}^*$. But the same X can always be regarded as the dual of some Y. As a member of a space of linear functionals on Y, it is an element of $\widehat{\mathbb{C}} \otimes Y^*$. But we can take $Y = X^*$, so that the element is a member of $\widehat{\mathbb{C}} \otimes X$. As an element of $X \otimes \widehat{\mathbb{C}}^*$ it can be written in the form $\psi \otimes \langle 1|$, while as an element of $\widehat{\mathbb{C}} \otimes X$ it will be written as $|1\rangle \otimes \psi$. In order to specify that ψ belongs to X, we write ψ^X.

A unitary transformation of W can be regarded as an element of $Hom\,(W, W)$, which is the standard realization of $W^* \otimes W$. A decomposable element of the latter space is written as

$$|x\rangle_{W^*} \otimes |y\rangle_W = |y\rangle \langle x| .$$

Hence, *a unitary transformation of* W *can be viewed as a ket of* $W^* \otimes W$. *The corresponding ray can be viewed as a state of the universe* $W^* \otimes W$. This state allows us to recover the action of the unitary transformation T on $\mathbb{P}(W)$. Indeed, this ray is the set $\mathbb{C}T$. If we impose the unitarity of $S = \lambda T$, we recover T up to a phase factor, which has no influence on its action on $\mathbb{P}(W)$.

A state of the universe W *can be represented through a rank one projector, which in turn can be regarded as a ket of* $W^* \otimes W$ *of the form* $|x\rangle_{W^*} \otimes |x\rangle_W$. This ket can be considered as a representative of a state from which it can be uniquely recovered. Indeed, if we impose the condition that λP is a projector, we obtain $\lambda = 1$. If a state of W is represented by a projector P of rank one, the transformed state under the unitary transformation T is TPT^*.

Let us discuss this relation in terms of tensor products. However, it is useful to discuss first the following problem. Let $\varphi \in Hom(K, L)$ and $\psi \in Hom(H, K)$. Then $\varphi \circ \psi \in Hom(H, L)$. The corresponding $|\varphi\rangle$ and $|\psi\rangle$ are kets of $K^* \otimes L$ and $H^* \otimes K$, respectively. The ket $|\varphi\rangle \otimes |\psi\rangle$ is an element of $(K^* \otimes L) \otimes (H^* \otimes K)$. Using the associative and the commutative property of the tensor products, we have the canonical isomorphism

$$\left(K^* \otimes L\right) \otimes \left(H^* \otimes K\right) \xrightarrow{\iota} \left(H^* \otimes L\right) \otimes \left(K^* \otimes K\right).$$

But the latter is nothing but $Hom(H, L) \otimes Hom(K, K)$. In the latter space, we define $\widehat{tr}_K (A \otimes B) = A\, tr\, B$ and extend it by bilinearity.

Finally, for any $|x\rangle \in (K^* \otimes L) \otimes (H^* \otimes K)$ we define $tr_K\, |x\rangle = \widehat{tr}_K\, \iota\,(|x\rangle)$. It is easy to show that $\varphi \circ \psi = tr_K\, |\varphi\rangle \otimes |\psi\rangle$.

We can put the matter in other terms. We consider a tensor product of the form $(K^* \otimes L) \otimes (H^* \otimes K)$ and a separable element in it admitting a factorization $|\psi\rangle \otimes |\varphi\rangle$ with $|\psi\rangle \in H^* \otimes K$ and $|\varphi\rangle \in K^* \otimes L$.

The transformation T corresponds to a ket $|T\rangle_{W^* \otimes W}$. We emphasize that, being the transformation a mapping, the rays involved in it must be regarded as variables, namely, as *different variables*, although taking values in the same set. The formalism of bra and kets allows to take into account this difference. Consider the variable ψ for a state of W. We can assign to it as values the rays of W. We will interpret this variable as an *initial* state of the system. If $|\psi\rangle$ is a normalized ket representing the ray, the equivalence class of normalized kets representing it will be denoted as $\psi\rangle$, that is, $\psi\rangle = \left\{e^{i\varphi}\,|\psi\rangle\right\}$. The variable ψ as a *final* state will be represented by $\langle\psi = \left\{e^{i\varphi}\,\langle\psi|\right\}$. In the usual notation, using kets to represent the states and primes to represent states after the transformation, we write $\psi' = T\psi$. Using representatives, we have $\left|\psi'\right\rangle = T\,|\psi\rangle$.

If no structure is imposed on H besides that of a Hilbert space, the group of allowed transformations is the full unitary group $\mathfrak{U}(H)$. In this case, the state of H is specified by a ray of H. If ψ denotes a state, we will denote $|\psi\rangle$ any ket representing it. As a rule, we will use normalized kets. With this condition, a ket representing a given state is defined up to an arbitrary phase factor. The state of the system may be subjected to a change. By this we mean that we are considering two states: a state ψ "before" the change and a state ψ' "after" the change. We emphasize that the terms "before" and "after" are rather conventional: no temporal evolution is in general understood.

The action of the group $\mathfrak{U}(H)$ passes to the space $\mathbb{P}(H)$ of the rays of H, and this latter action is transitive. For notational simplicity, we will denote with the same symbol the action of an element of $\mathfrak{U}(H)$ on H and on $\mathbb{P}(H)$. For every pair $(\psi, \psi') \in \mathbb{P}(H)$, the set of elements $T \in \mathfrak{U}(H)$ such that $\psi' = T\psi$ is nonempty: it consists of a left coset of the stabilizer of ψ.

We can "explain" why a state of a universe is represented by a ray and not by a ket in the following way. Consider the tensor product $\widehat{\mathbb{C}} \otimes H$ where $\widehat{\mathbb{C}}$ is $\mathbb{C}$ regarded as a vector space over itself with the structure of Hilbert space defined by the scalar product $\langle x | y \rangle = x^* y$. All of its elements are separable, and the position $|c\rangle \otimes |x\rangle \rightarrow c\,|x\rangle$ defines an isomorphism between $\widehat{\mathbb{C}} \otimes H$ and H.

We associate to a universe W an extended system that we call extended universe W_e. If W is the Hilbert space of W, we introduce for W_e the Hilbert space $W_e = \widehat{\mathbb{C}} \otimes W$. The states of an extended universe are kets in its Hilbert space W_e. By definition, in an extended universe, different kets correspond to different states. We assume that a state $|w_e\rangle$ of the extended universe defines uniquely the state of the corresponding universe. A state of W_e can be always be represented as $|c\rangle \otimes |x\rangle$, but each factor is defined up to a scalar coefficient.

Chapter 2
The State of a Quantum System as a Subsystem of a Composite System

An important difference between the notion of a state of a physical system in quantum and in classical mechanics is the fact that, while in the latter the nature of a state of a subsystem of a given system is the same as that of the whole system, in quantum mechanics their character is deeply different. Indeed, the presence of entanglement for the states of composite systems prevents, in general, the possibility of ascribing to a subsystem a definite state, in the sense of a pure state of an isolated system. In the conventional setting of the interpretative rules of quantum mechanics, the state of the subsystem is described by a partial density operator, whose definition is justified by the probabilistic interpretation and the use of Born's rule.

2.1 Non-probabilistic Characterization of the State of a Subsystem

In this section, we wish to arrive at a characterization of the state of a subsystem of a quantum system independently from any probabilistic interpretation. Starting from a natural way of defining such a state, we will find that one way of characterizing it is through a partial density operator *independently of any a priori probabilistic interpretation*. The latter will come out as a consequence of the formalism and of a few reasonable assumptions, connected with the notion of a state.

Suppose that a composite system $\mathscr{C}$ consists of two parts $\mathscr{S}$ (the "system") and $\mathscr{E}$ (the "environment"). We denote with $\mathscr{H}$ the Hilbert space of $\mathscr{S}$ and with $\mathscr{K}$ the Hilbert space of the environment. As the goal of this work is merely an assessment

The content of this chapter can be found in Campanella's files *ZurekState* (2011-09-01) and *sav-Consideration on Zurek's interpretation of quantum mechanics* (2010-09-05).

M. Campanella et al., *Interpretative Aspects of Quantum Mechanics*, UNIPA Springer Series,
https://doi.org/10.1007/978-3-030-44207-1_2

of interpretative aspects of Quantum Mechanics, we will avoid the technicalities connected with infinite-dimensional Hilbert spaces, so that all the Hilbert spaces involved in our discussion will be supposed *finite-dimensional*. It is hoped that the generalization to the infinite-dimensional case does not involve significant changes of the *interpretative framework*.

A key step for the introduction of a natural definition of a state of $\mathscr{S}$ as a subsystem of the composite system is the formalization of a property indicated by Zurek. In the words of Zurek: "Unitary transformations must act on the system to alter its state. That is, when an operator does not act on the Hilbert space of $\mathscr{S}$, i.e., when it has the form $I \otimes (\ )$, the state of $\mathscr{S}$ does not change." If we agree with this assertion, and we describe the states of the composite system as pure states (according to the idea that the composite system is "the Universe"), we conclude that the state of $\mathscr{S}$ associated with a pure state $|\psi\rangle$ of the composite system must depend only on the orbit of $|\psi\rangle$ under the action of the group $I \otimes \mathscr{U}(\mathscr{K})$. *We will suppose conversely that different orbits correspond to different states*. This is not a strong assumption as it may appear at first sight. Indeed, if there were different orbits physically indistinguishable, we could always redefine the states through the passage to suitable equivalence classes.

In this way, we may tentatively define the state of $\mathscr{S}$ as a subsystem of the composite system associated with $|\psi\rangle$ simply as the orbit of $|\psi\rangle$. According to this definition, the state of $\mathscr{S}$ is characterized as a subset of the space of the composite system. Instead, we would like to arrive at a definition which involves only points of the Hilbert space of $\mathscr{S}$. To this purpose, we observe that any set of mathematical objects in bijection with the orbits of $I \otimes \mathscr{U}(\mathscr{K})$ can be as well used to define the state.

If we adopt this (implicit) definition of a state of $\mathscr{S}$ in the presence of the environment $\mathscr{E}$, we can arrive in a natural way at a characterization of a state of $\mathscr{S}$ as a density operator. We emphasize that in this association no a priori probabilistic interpretation (and, a fortiori, no Born's rule) will be involved. Indeed, the density operator will arise as an *orbit invariant*.

2.2 Mathematical Characterization of Zurek's Envariance

A central concept in Zurek's approach to quantum mechanics is envariance (environment-induced superselection). Let us consider a quantum system $\mathscr{S}$ interacting with another system, which will be called the environment and will be denoted $\mathscr{E}$. A state of the composite system is the ray defined by a vector $|\psi\rangle$ of the tensor product $\mathscr{H} \otimes \mathscr{K}$ of the Hilbert spaces of $\mathscr{S}$ and $\mathscr{E}$, respectively. The vector $|\psi\rangle$ can be expanded in a Schmidt basis as follows:

$$|\psi\rangle = \sum a_k |s_k\rangle |\varepsilon_k\rangle, \tag{2.2.1}$$

where $|s_k\rangle$ and $|\varepsilon_k\rangle$ form orthonormal bases of $\mathscr{H}$ and $\mathscr{K}$, respectively.

According to Zurek, we say that $|\psi\rangle$ is invariant with respect to $U \in \mathscr{U}(\mathscr{H})$ if there is $V \in \mathscr{U}(\mathscr{K})$ such that $U \otimes V|\psi\rangle = |\psi\rangle$.

It is easy to show that from the following theorem.

Theorem 2.1 *The set of transformations under which $|\psi\rangle$ is invariant is a group.*

Indeed, let S be the set of such transformations. If $U \in S$, there is $V \in \mathscr{U}(\mathscr{K})$ such that

$$U \otimes V|\psi\rangle = |\psi\rangle.$$

Therefore,

$$U^{-1} \otimes V^{-1}|\psi\rangle = |\psi\rangle,$$

so that $U^{-1} \in S$. Furthermore, if $U_1 \in S$ and $U_2 \in S$, there are $V_1, V_2 \in \mathscr{U}(\mathscr{K})$ such that

$$U_1 \otimes V_1|\psi\rangle = |\psi\rangle \qquad \text{and} \qquad U_2 \otimes V_2|\psi\rangle = |\psi\rangle.$$

Therefore, $U_1U_2 \otimes V_1V_2|\psi\rangle = |\psi\rangle$ and $U_1U_2 \in S$. □

In what follows $tr_S(\ )$ and $tr_E(\ )$ will denote the partial traces over $\mathscr{S}$ and $\mathscr{E}$, respectively. We now prove the following.

Theorem 2.2 *The state $|\psi\rangle$ is invariant under $U \in \mathscr{U}(H)$ if and only if*

$$U\, tr_E(|\psi\rangle\langle\psi|)\, U^{-1} = tr_E(|\psi\rangle\langle\psi|).$$

Proof If $|\psi\rangle$ is invariant under U, there is $V \in \mathscr{U}(K)$ such that $U \otimes V|\psi\rangle = |\psi\rangle$. Using (2.2.1), we get

$$|\psi\rangle\langle\phi| = \sum a_k a_h^* \, |s_k \varepsilon_k\rangle\langle s_h \varepsilon_h|.$$

Therefore

$$U \otimes V|\psi\rangle = |\psi\rangle\langle\psi|\,(U \otimes V)^* = \sum a_k a_h^* U|s_k\rangle\langle s_h|U^* \otimes V|\varepsilon_h\rangle\langle\varepsilon_k|V^*$$

and

$$tr_E|\psi\rangle\langle\psi| = \sum a_k a_h^* tr_E|s_k\rangle\langle s_h| \otimes |\varepsilon_h\rangle\langle\varepsilon_k| = \sum |a_k|^2|s_k\rangle\langle s_k|.$$

Similarly,

$$tr_E U \otimes V|\psi\rangle\langle\psi|(U \otimes V)^* = \sum a_k a_h^* U|s_k\rangle\langle s_h|U^* tr(V|\varepsilon_h\rangle\langle\varepsilon_k|V^*).$$

But

$$tr(V|\varepsilon_h\rangle\langle\varepsilon_k|V^*) = \langle\varepsilon_h|\varepsilon_k\rangle = \delta_{hk},$$

therefore

$$tr_E U \otimes V|\psi\rangle\langle\psi|(U \otimes V)^* = \sum |a_k|^2 U|s_k\rangle\langle s_k|U^*.$$

But

$$U \otimes V|\psi\rangle = |\psi\rangle,$$

so that

$$tr_E|\psi\rangle\langle\psi| = U \sum |a_k|^2|s_k\rangle\langle s_k|U^* = U tr_E|\psi\rangle\langle\psi|\, U^*.$$

Conversely, suppose

$$U\, tr_E(|\psi\rangle\langle\psi|)\, U^{-1} = tr_E(|\psi\rangle\langle\psi|).$$

The operator $\rho = tr_E|\psi\rangle\langle\psi|$ is (Hermitian) positive. Therefore, we have the spectral decomposition $\rho = \sum p_k P_k$ with $p_k \geq 0$. If

$$|\psi\rangle = \sum a_k|r_k\rangle|\varepsilon_k\rangle \tag{2.2.2}$$

is a Schmidt decomposition of $|\psi\rangle$ we have $\rho = \sum |a_l|^2|r_l\rangle\langle r_l|$. We can put together the terms with equal coefficients getting $\rho = \sum_k g_k \sum_l |r_{kl}\rangle\langle r_{kl}|$. Consequently, we get $g_k = p_k$ and $P_k = \sum_l |r_{kl}\rangle\langle r_{kl}|$. Similarly

$$U \otimes I|\psi\rangle = \sum a_k|r_k'\rangle|\varepsilon_k\rangle, \tag{2.2.3}$$

with $|r_k'\rangle = U|r_k\rangle$ and

$$tr_E(U \otimes I)|\psi\rangle\langle\psi|(U \otimes I)^* = \sum |a_k|^2|r_k'\rangle\langle r_k'| = U tr|\psi\rangle\langle\psi|U^*.$$

Therefore

$$\sum p_k P_k = \sum |a_l|^2|r_l'\rangle\langle r_l'|.$$

Putting together terms with equal coefficients, we get

$$\sum p_k P_k = \sum_k p_k \sum_l |a_l|^2|r_{kl}'\rangle\langle r_{kl}'|, \tag{2.2.4}$$

therefore $P_k = \sum_l |a_l|^2|r_{kl}'\rangle\langle r_{kl}'|$. This means that $\{|r_{kl}\rangle\}$ and $\left\{|r_{kl}'\rangle\right\}$ are orthonormal bases of the same space. We can then write

$$|r_{kt}'\rangle = \sum_l \tau_{lt}^{(k)}|r_{kl}\rangle,$$

where $\tau_{lt}^{(k)}$ are the coefficients of a unitary matrix. Regrouping terms in (2.2.3) and in (2.2.2) in the same way as in (2.2.4), we get

$$U \otimes I|\psi\rangle = \sum_{kt} a_{kt}|r'_{kt}\rangle|\varepsilon_{kt}\rangle \qquad \text{and} \qquad |\psi\rangle = \sum_{kt} a_{kl}|r_{kt}\rangle|\varepsilon_{kt}\rangle.$$

Consequently, we get

$$U \otimes I|\psi\rangle = \sum_{kt} a_{kt} \sum_{l} \tau_{lt}^{(k)}|r_{kl}\rangle|\varepsilon_{kt}\rangle.$$

But

$$\sum_{kt} a_{kt} \sum_{l} \tau_{lt}^{(k)}|r_{kl}\rangle|\varepsilon_{kt}\rangle = \sum_{kt} |r_{kl}\rangle \sum_{l} a_{kt}\tau_{lt}^{(k)}|\varepsilon_{kt}\rangle$$

and

$$a_{kt} = p_k^{1/2} e^{i\vartheta_{kt}},$$

so that

$$U \otimes I|\psi\rangle = \sum_{kl} |r_{kl}\rangle \sum_{t} p_k^{1/2} e^{i\vartheta_{kt}} \tau_{lt}^{(k)}|\varepsilon_{kt}\rangle.$$

We have

$$|\psi\rangle = \sum_{kl} p_k^{1/2}|r_{kl}\rangle e^{i\vartheta_{kl}}|\varepsilon_{kl}\rangle \qquad \text{and} \qquad U \otimes I|\psi\rangle = \sum_{kl} p_k^{1/2}|r_{kl}\rangle \sum_{t} e^{i\vartheta_{kt}}|\varepsilon_{kt}\rangle.$$

Putting $e^{i\vartheta_{kl}}|\overline{\varepsilon}_{kl}\rangle = \sum_t e^{i\vartheta_{kt}}|\varepsilon_{kt}\rangle$, we get

$$U \otimes I|\psi\rangle = \sum_{kl} p_k^{1/2}|r_{kl}\rangle e^{i\vartheta_{kl}}|\overline{\varepsilon}_{kl}\rangle.$$

From the definition, we see immediately that $|\overline{\varepsilon}_{kl}\rangle = V^*|\varepsilon_{kl}\rangle$ for some $V \in \mathscr{U}(\mathscr{K})$. Therefore, we have

$$U \otimes I|\psi\rangle = I \otimes V^*|\psi\rangle,$$

whence the thesis. □

We can state the above result in the equivalent form.

Theorem 2.3 *The group of envariance of a state of the composite system in $\mathscr{S}$ is the group of invariance of the corresponding density operator in $\mathscr{S}$.*

Of course the role of $\mathscr{S}$ and $\mathscr{E}$ can be exchanged, so that the group of envariance of a state of the composite system in $\mathscr{E}$ is the group of invariance of the corresponding density operator in $\mathscr{S}$.

We first recall (see Appendix D) that, in the finite-dimensional case, there is a canonical isomorphism between $\mathscr{H} \otimes \mathscr{K}$ and the Hilbert space $Hom(\mathscr{H}^*, \mathscr{K})$ of all linear maps from $\mathscr{H}^*$ to $\mathscr{K}$, equipped with the scalar product $\langle\psi|\psi\rangle = tr(\psi^*\psi)$. In this isomorphism, the corresponding of the ket $|\psi\rangle \in \mathscr{H} \otimes \mathscr{K}$ will be denoted

ψ. The action of $I \otimes V$ ($V \in \mathscr{U}(\mathscr{K})$) on $|\psi\rangle$ corresponds to the action $\psi \mapsto V\psi$. The following theorem holds.

Theorem 2.4 *The necessary and sufficient condition for the existence of a transformation $V \in \mathscr{U}(\mathscr{K})$ such that $\psi' = V\psi$ is that $\psi'^*\psi' = \psi^*\psi$.*

For a proof of this theorem, see Appendix D.

Owing to the above theorem, the state of $\mathscr{S}$ corresponding to $|\psi\rangle$ can be characterized by $\psi^*\psi$. The latter is a positive self-adjoint operator operating on $\mathscr{H}^*$ on the left if $\mathscr{H}^*$ is regarded as a space of kets, and on the right if $\mathscr{H}^*$ is regarded as the space of bras associated with $\mathscr{H}$, and hence it operates on the left on $\mathscr{H}$. Furthermore, if $|\psi\rangle$ is normalized, $tr\psi^*\psi = 1$, so that $\psi^*\psi$ is a density operator. It easy is to prove that

$$\psi^*\psi = tr_E\,|\psi\rangle\,\langle\psi|\,. \tag{2.2.5}$$

Indeed, let

$$|\psi\rangle = \sum a_k\,|s_k\rangle\,|\varepsilon_k\rangle \tag{2.2.6}$$

be a Schmidt decomposition of $|\psi\rangle$. Denoting $|x\rangle^*$a bra of $\mathscr{H}$ regarded as a ket of $\mathscr{H}^*$, the corresponding linear map ψ is

$$\psi = \sum a_k\,\big|\varepsilon_k\rangle(^*\langle s_k\big|), \tag{2.2.7}$$

so that

$$\psi^*\psi = \sum |a_k|^2 |s_k\rangle^{**}\,\langle s_k|. \tag{2.2.8}$$

If $\langle x|$ is a bra of $\mathscr{H}$, it must be written $|x\rangle^*$ when regarded as a ket of $\mathscr{H}^*$; we can then write

$$\psi^*\psi\,|x\rangle^* = \sum |a_k|^2 |s_k\rangle^{**}\langle s_k|x\rangle^* = \sum |a_k|^2 |s_k\rangle^*\,\langle x|s_k\rangle. \tag{2.2.9}$$

Regarding the latter as a bra of $\mathscr{H}$, we get

$$\langle x|\,\psi^*\psi = \langle x|\sum |a_k|^2\,|s_k\rangle\,\langle s_k|. \tag{2.2.10}$$

So that

$$\psi^*\psi = \sum |a_k|^2\,|s_k\rangle\,\langle s_k|. \tag{2.2.11}$$

On the other hand, we have

$$|\psi\rangle\,\langle\psi| = \sum a_k a_h^*\,|s_k\rangle\,\langle s_h| \otimes |\varepsilon_k\rangle\,\langle\varepsilon_h|, \tag{2.2.12}$$

and $tr_E\,|\psi\rangle\,\langle\psi| = \sum |a_k|^2\,|s_k\rangle\,\langle s_k| = \psi^*\psi$.

The following theorem holds.

Theorem 2.5 *If $\psi \in Hom(\mathcal{H}^*, \mathcal{K})$, there is a unique orthogonal decomposition $\mathcal{H} = \oplus \mathcal{H}_a$ such that the restriction of ψ to each $\mathcal{H}_a^*$ is a dilatation on its image and that images corresponding to different subspaces are orthogonal. The subspaces $\mathcal{H}_a$ are the eigenspaces of $\psi^*\psi$ and the corresponding dilatation coefficients are the square roots of the associated eigenvalues.*

Proof Let H_α^* be the eigenspace of $\psi^*\psi$ associated with the eigenvalue p_α. If $\langle x| \in H_\alpha^*$, it can be regarded as a ket of $\mathcal{H}^*$ and we can write it as $|x\rangle^*$. If $|y\rangle = \psi|x\rangle^*$, we have

$$\langle y|y\rangle = {}^*\langle x|\psi^*\psi|x\rangle^* = p_\alpha^*\langle x|x\rangle.$$

Therefore, the restriction of ψ to H_α^* is a dilatation, and the dilatation coefficient is $\sqrt{p_\alpha}$. The orthogonal decomposition $\mathcal{H} = \oplus H_\alpha$ is such that the restriction of ψ to H_α^* is a dilatation. Furthermore, if $|x\rangle^* \in H_\alpha^*$ and $|x'\rangle^* \in H_\beta^*$ $(\alpha \neq \beta)$,

$${}^*\langle x|\psi^*\psi|x\rangle^* = p_\alpha^*\langle x|x\rangle = 0,$$

so that the images of H_α^* and H_β^* are orthogonal. Conversely, let $\mathcal{H} = \oplus H_\alpha$ be an orthogonal decomposition such that the restriction of ψ to each H_α^* is a dilatation and the images of different subspaces are orthogonal. We have the orthogonal decomposition $\mathcal{H}^* = \oplus H_\alpha^*$. If $\{P_\alpha\}$ is the associated set of projectors, we have

$${}^*\langle x|\psi^*\psi|x\rangle^* = \sum {}^*\langle x|P_\alpha\psi^*\psi P_\beta|x\rangle^*.$$

As the images of different subspaces are orthogonal, we get

$${}^*\langle x|\psi^*\psi|x\rangle^* = \sum {}^*\langle x|P_\alpha\psi^*\psi P_\alpha|x\rangle^* = \sum d_\alpha^{2*}\langle x|P_\alpha P_\alpha|x\rangle^* = {}^*\langle x|\sum d_\alpha^2 P_\alpha|x\rangle^*,$$

where d_α are the dilatation coefficients.

We conclude that $\psi^*\psi = \sum d_\alpha^2 P_\alpha$ and we recognize in this formula the spectral decomposition of $\psi^*\psi$. Therefore, the H_α^* are the eigenspaces of $\psi^*\psi$ and the decomposition is unique. In this unique decomposition, the subspaces are the eigenspaces of $\psi^*\psi$ and the dilatation coefficients are the square roots of the eigenvalues. □

For each α there is a unique linear mapping ψ_α such that $\psi_\alpha|x\rangle^* = \psi|x\rangle^*$ for $|x\rangle^* \in H_\alpha^*$ and for $\psi_\alpha|x\rangle^* = 0$ for $|x\rangle^* \in H_{\alpha\perp}^*$. Of course, we have

$$\phi = \sum \psi_\alpha.$$

Furthermore, the image of ψ_α is the same as the image of the restriction of ψ to H_α^*. Let us consider ${}^*\langle x|\psi_\beta^*\psi_\alpha|x'\rangle^*$ for $\beta \neq \alpha$. It is certainly zero for $|x'\rangle^* \in H_\gamma^*$ with $\gamma \neq \alpha$ and for $|x'\rangle^* \in H_\delta^*$ with $\delta \neq \beta$. If $|x\rangle^* \in H_\beta^*$ and $|x'\rangle^* \in H_\alpha^*$, $\psi_\alpha|x'\rangle^*$ is in the image of the restriction of ψ to H_α^* and $\psi_\beta|x\rangle^*$ is in the image of the restriction of ψ to H_β^*. Therefore, ${}^*\langle x|\psi_\beta^*\psi_\alpha|x'\rangle^* = 0$ when $|x\rangle^*$ and $|x'\rangle^*$ lie in any of the spaces H_γ^*. By bilinearity, we conclude that $\psi_\beta^*\psi_\alpha = 0$. Let us now consider

${}^*\langle x|\psi_\alpha^*\psi_\alpha|x\rangle^*$. It is certainly zero when $|x\rangle^*$ belongs to some H_γ^* with $\gamma \neq \alpha$. If $|x\rangle^* \in H_\gamma^*$, ${}^*\langle x|\psi_\alpha^*\psi_\alpha|x\rangle^* = p_\alpha^{**}\langle x|x\rangle^*$. We conclude that $\psi_\alpha^*\psi_\alpha = p_\alpha P_\alpha$. From the above considerations, the following theorem is derived.

Theorem 2.6 *If $\psi \in Hom\,(H, K)$ is nonzero, there is a unique decomposition*

$$\psi = \sum d_\alpha \chi_\alpha$$

such that the d_α are positive coefficients all different from each other, $\chi_\alpha^*\chi_\beta = 0$ for $\alpha \neq \beta$, and $\{\chi_\alpha^*\chi_\alpha\}$ is a family of orthogonal projectors.

Indeed, it is sufficient to associate to each $p_\alpha \neq 0$ the mapping $\chi_\alpha = p_\alpha^{-1/2}\psi_\alpha$ and $d_\alpha = p_\alpha^{1/2}$.

The following theorem, which can be regarded as a kind of Schmidt decomposition theorem in an invariant form, will be useful in subsequent developments.

Theorem 2.7 *If $|\psi\rangle \in \mathscr{H} \otimes \mathscr{K}$, there is a unique decomposition*

$$|\psi\rangle = \sum d_\alpha\,|\chi_\alpha\rangle$$

such that $d_\alpha > 0$ with the d_α all different from each other, $tr_E\,|\chi_\alpha\rangle\langle\chi_\beta| = 0$ for $\alpha \neq \beta$, and the set $\{P_\alpha = tr_E\,|\chi_\alpha\rangle\,\langle\chi_\alpha|\}$ is a family of orthogonal projectors of $\mathscr{H}$.

Proof In the canonical isomorphism between $\mathscr{H} \otimes \mathscr{K}$ and $Hom(\mathscr{H}^*, \mathscr{K})$, the corresponding of $\varphi \in Hom(\mathscr{H}^*, \mathscr{K})$ is the ket $|\varphi\rangle$: if $\varphi = |v\rangle^*\langle u|$, then $|\varphi\rangle = |u\rangle|v\rangle$; furthermore, $\varphi^* = |u\rangle^*\langle v|$. Therefore, if $\psi = |v'\rangle^*\langle u'|$, then $\psi^*\varphi = |u'\rangle^*\langle v'|v\rangle^*\langle u|$. On the other hand, we have $|\psi\rangle = |u'\rangle|v'\rangle$ and $|\varphi\rangle\langle\psi| = |u\rangle\langle u'| \otimes |v\rangle\langle v'|$. We have $tr_E|\varphi\rangle\langle\psi| = |u\rangle\langle v'|v\rangle\langle u'|$. But $|u'\rangle^* = \langle u'|$ and $\langle u| = |u\rangle^*$ so that

$$\psi^*\varphi = tr_E|\varphi\rangle\langle\psi|. \tag{2.2.13}$$

By linearity this equation must hold for arbitrary φ and ψ. Hence, Theorem 2.7 follows from Theorem 2.6 as a consequence of the canonical isomorphism.

Henceforth, the decomposition introduced in the above theorem will be called the *canonical decomposition*. This decomposition can be regarded as the abstract counterpart of the well-known singular value decomposition of a matrix.

Remark 2.1 The kets $|\chi_\alpha\rangle$ in the canonical decomposition are not normalized in general. Indeed, $\langle\chi_\alpha|\chi_\alpha\rangle = tr\,|\chi_\alpha\rangle\,\langle\chi_\alpha| = tr_E\,|\chi_\alpha\rangle\,\langle\chi_\alpha| = n_\alpha$ where n_α is the dimension of the projection space of P_α. Sometimes it is convenient to recast the canonical decomposition in a form involving normalized kets. To this purpose we put $|\chi'_\alpha\rangle = 1/\sqrt{n_\alpha}\;|\chi_\alpha\rangle$, so that $|\psi\rangle = \sum d'_\alpha\,|\chi'_\alpha\rangle$ with $d'_\alpha = \sqrt{n_\alpha}d_\alpha$. The ket $|\psi\rangle$ is normalized whenever $\sum(d'_\alpha)^2 = 1$. The associated density operator is $\rho_S = \sum d_\alpha^2 P_\alpha$. Hence, the normalization condition for ρ_S is $\sum n_\alpha d_\alpha^2 = 1$.

2.3 Characterization of a State Through Its Properties

We know that a state ζ of an isolated system described by a Hilbert space $\mathscr{X}$ is characterized by the ray associated with some $|\psi\rangle$, i.e., by the set $\{\lambda\,|\psi\rangle : \lambda \in \mathbb{C}\}$ which is a point of the projective space $\mathbb{P}(\mathscr{X})$. The unitary group $\mathscr{U}(\mathscr{X})$ acts naturally on the rays and hence on $\mathbb{P}(\mathscr{X})$. However, the action is not faithful, while the action of the quotient $\mathscr{U}^Z(\mathscr{X})$ of $\mathscr{U}(\mathscr{X})$ modulo its center is. Therefore, we will take $\mathscr{U}(\mathscr{X})$ as the structure group of $\mathbb{P}(\mathscr{X})$ (in general we will denote G^Z the quotient group of G modulo its center).

We say that a property $\mathfrak{P}(x)$ of the nonzero kets of $\mathscr{X}$ is a property of the states of the system if, whenever $\mathfrak{P}$ is true for a nonzero ket $|x\rangle$, it is true for all the kets of the ray generated by $|x\rangle$. We say that a property $\mathfrak{D}(x)$ of the states of the system defines a state of the system if, for every property $\mathfrak{P}(x)$ of the states of the system, the condition $\mathfrak{P}(x) \Rightarrow \mathfrak{D}(x)$ entails $\mathfrak{P}(x) \Leftrightarrow \mathfrak{D}(x)$.

We say that a state is completely defined if it is constrained by a property of the type $\mathfrak{D}(x)$. We say that the state is partially defined if it is constrained by a weaker property of the states of the system.

The systems $\mathscr{S}$ and $\mathscr{E}$ are represented by $\mathbb{P}(\mathscr{H})$ and $\mathbb{P}(\mathscr{K})$, respectively, while the composite system $\mathscr{C}$ is represented by $\mathbb{P}(\mathscr{H} \otimes \mathscr{K})$. Their structure groups are $\mathscr{U}^Z(\mathscr{H})$, $\mathscr{U}^Z(\mathscr{K})$, and $\mathscr{U}^Z(\mathscr{H} \otimes \mathscr{K})$. The elements of $\mathscr{U}(\mathscr{H} \otimes \mathscr{K})$ acting on indecomposable kets as $|h\rangle\,|k\rangle \mapsto |h\rangle\, V\,|k\rangle$ with $V \in \mathscr{U}(\mathscr{K})$ will be denoted $I \otimes V$ and form a subgroup $Id_{\mathscr{H}} \otimes \mathscr{U}(\mathscr{K})$ of $\mathscr{U}(\mathscr{H} \otimes \mathscr{K})$.

We say that a property $\mathfrak{P}(x)$ of the nonzero kets of $\mathbb{P}(\mathscr{H} \otimes \mathscr{K})$ is a property of the states of the subsystem $\mathscr{S}$ of $\mathscr{C}$ if it is a property of the states of $\mathscr{C}$ and, whenever $\mathfrak{P}$ is true for a ket $|\psi\rangle$, it is true for all the kets of the form $I \otimes V\,|\psi\rangle$ with $V \in \mathscr{U}(\mathscr{K})$.

Let Ω be a nonzero orbit of $Id_{\mathscr{H}} \otimes \mathscr{U}(\mathscr{K})$, and $\widehat{\Omega} = \mathbb{C}\Omega$. Then the property $\mathfrak{P}_{\widehat{\Omega}}(x)$: " $\mathfrak{P}_{\widehat{\Omega}}(x)$ is true iff $x \in \widehat{\Omega}$" is a property of the states of the subsystem $\mathscr{S}$ of $\mathscr{C}$. The property just defined satisfies the condition: if $\mathfrak{P}'(x)$ is a property of the states of the subsystem $\mathscr{S}$ of $\mathscr{C}$ such that $\mathfrak{P}'(x) \Rightarrow \mathfrak{P}_{\widehat{\Omega}}(x)$, then $\mathfrak{P}'(x) \Leftrightarrow \mathfrak{P}_{\widehat{\Omega}}(x)$.

We say that a property $\mathfrak{D}(x)$ of the states of the subsystem $\mathscr{S}$ of $\mathscr{C}$ defines a state of the subsystem $\mathscr{S}$ of $\mathscr{C}$ if whenever a property $\mathfrak{P}(x)$ of the states of the subsystem $\mathscr{S}$ of $\mathscr{C}$ is such that $\mathfrak{P}(x) \Rightarrow \mathfrak{D}(x)$ then $\mathfrak{P}(x) \Leftrightarrow \mathfrak{D}(x)$.

We say that $\mathfrak{D}(x)$ and $\mathfrak{D}'(x)$ define the same state of the subsystem $\mathscr{S}$ of $\mathscr{C}$ if $\mathfrak{D}(x) \Leftrightarrow \mathfrak{D}'(x)$. If $\widehat{\Omega} = \mathbb{C}\Omega$, $\mathfrak{P}_{\widehat{\Omega}}(x)$ defines a state of the subsystem $\mathscr{S}$ of $\mathscr{C}$. If $\mathfrak{D}(x)$ defines a state of the subsystem $\mathscr{S}$ of $\mathscr{C}$ there is a $\widehat{\Omega} = \mathbb{C}\Omega$ such that $\mathfrak{D}(x) \Leftrightarrow \mathfrak{P}_{\widehat{\Omega}}(x)$. Conversely, if there is a $\widehat{\Omega} = \mathbb{C}\Omega$ such that $\mathfrak{D}(x) \Leftrightarrow \mathfrak{P}_{\Xi}(x)$, $\mathfrak{D}(x)$ defines a state of the subsystem $\mathscr{S}$ of $\mathscr{C}$. Of course $\mathfrak{P}_{\widehat{\Omega}}(x) \Leftrightarrow \mathfrak{P}_{\widehat{\Omega}'}(x)$ iff $\widehat{\Omega} = \widehat{\Omega}'$.

Let $\mathfrak{D}(x)$ be a property of the states of the composite system $\mathscr{C}$. The following theorem holds.

Theorem 2.8 $\mathfrak{D}(x)$ *defines a state of the subsystem* $\mathscr{S}$ *of* $\mathscr{C}$ *iff there is a density operator* ρ_S *such that* $\mathfrak{D}(x) \Leftrightarrow \langle x|x\rangle^{-1} tr_E\,|x\rangle\,\langle x| = \rho_S$.

$\langle x|x\rangle^{-1} tr_E |x\rangle\langle x| = \rho_S$ and $\langle x|x\rangle^{-1} tr_E |x\rangle\langle x| = \rho'_S$ define the same state of the subsystem $\mathscr{S}$ of $\mathscr{C}$ iff $\rho_S = \rho'_S$. If Dim $\mathscr{H} \leq$ Dim $\mathscr{K}$*, for every density operator ρ_S, $\langle x|x\rangle^{-1} tr_E |x\rangle\langle x| = \rho_S$ defines a state of the subsystem $\mathscr{S}$ of $\mathscr{C}$.*

Proof If $\mathfrak{D}(x)$ defines a state of the subsystem $\mathscr{S}$ of $\mathscr{C}$ there is a unique $\widehat{\Omega} = \mathbb{C}\Omega \neq 0$ (being Ω an orbit of $I \otimes \mathscr{U}(\mathscr{K})$) such that $\mathfrak{D}(x)$ is true whenever $x \in \widehat{\Omega}$. Let $|\psi\rangle$ be a nonzero ket of $\widehat{\Omega}$. Then $x \in \widehat{\Omega}$ iff there are $\lambda \in \mathbb{C}$ $(\lambda \neq 0)$ and $|x'\rangle$ such that $|x'\rangle$ belongs to the orbit of $|\psi\rangle$ and $|x'\rangle = \lambda |x\rangle$. We define

$$\rho_S = \langle\psi|\psi\rangle^{-1} tr_E |\psi\rangle\langle\psi| .$$

Then $|x'\rangle$ belongs to the orbit of $|\psi\rangle$ iff $tr_E |x'\rangle\langle x'| = \langle\psi|\psi\rangle \rho_S$, i.e., iff

$$|\lambda|^2 tr_E |x\rangle\langle x| = \langle\psi|\psi\rangle \rho_S.$$

Taking the traces with respect to $\mathscr{S}$ of both sides we get $|\lambda|^2 = \langle x|x\rangle^{-1} \langle\psi|\psi\rangle$ and hence $\langle x|x\rangle^{-1} tr_E |x\rangle\langle x| = \rho_S$. Then $\mathfrak{D}(x) \Leftrightarrow \langle x|x\rangle^{-1} tr_E |x\rangle\langle x| = \rho_S$.

Conversely, suppose that there is a density operator such that

$$\mathfrak{D}(x) \Leftrightarrow \langle x|x\rangle^{-1} tr_E |x\rangle\langle x| = \rho_S.$$

Let $\widehat{\Omega} = \{|x\rangle \,|\langle x|x\rangle^{-1} tr_E |x\rangle\langle x| = \rho_S\}$. Let $\mathfrak{D}(x)$ be true for $x = |\psi\rangle$. Therefore, $\rho_S = \langle\psi|\psi\rangle^{-1} tr_E |\psi\rangle\langle\psi|$ so that $|\psi\rangle \in \widehat{\Omega}$. Let Ω be the orbit of $|\psi\rangle$. $|x\rangle \in \widehat{\Omega}$ iff $\langle x|x\rangle^{-1} tr_E |x\rangle\langle x| = \langle\psi|\psi\rangle^{-1} tr_E |\psi\rangle\langle\psi|$. Putting $|\psi'\rangle = \langle x|x\rangle^{-1/2} \langle\psi|\psi\rangle^{1/2} |x\rangle$ we get $tr_E |\psi'\rangle\langle\psi'| = tr_E |\psi\rangle\langle\psi|$, so that $\widehat{\Omega} = \mathbb{C}\Omega$. Therefore, $\mathfrak{D}(x) \Leftrightarrow x \in \widehat{\Omega}$ and $\mathfrak{D}(x)$ defines a state of the subsystem $\mathscr{S}$ of $\mathscr{C}$. As $\widehat{\Omega} = \{|x\rangle \,|\langle x|x\rangle^{-1} tr_E |x\rangle\langle x| = \rho_S\}$ and $\widehat{\Omega}' = \{|x\rangle \,|\langle x|x\rangle^{-1} tr_E |x\rangle\langle x| = \rho'_S\}$ coincide iff $\rho_S = \rho'_S$, the states coincide iff the latter condition holds. Let ρ_S be a density operator. In order to show that it defines a state of the subsystem $\mathscr{S}$ of $\mathscr{C}$, it is sufficient to prove that the equation $\langle x|x\rangle^{-1} tr_E |x\rangle\langle x| = \rho_S$ has solutions. Starting from the spectral decomposition $\rho_S = \sum p_\alpha P_\alpha$ we can choose an orthonormal basis adapted to the decomposition and select a bijection between this basis and an orthonormal set of vectors of $\mathscr{K}$. If $|\chi_\alpha\rangle$ denotes the normalized sum of the tensor products of corresponding vectors associated with the eigenspace defined by P_α, the vector of $\mathscr{K}$ defined as $|\psi\rangle = \sum \sqrt{p_\alpha} |\chi_\alpha\rangle$ is normalized and $\rho_S = tr_E |\psi\rangle\langle\psi| = \langle\psi|\psi\rangle^{-1} tr_E |\psi\rangle \langle\psi|$. □

We observe that the states of the subsystem $\mathscr{S}$ of $\mathscr{C}$ have been defined indirectly, i.e., through the formulation according to which some suitable specific property of the nonzero kets of $\mathscr{C}$ defines a state of the subsystem $\mathscr{S}$ of $\mathscr{C}$. Among the infinite possibilities we can, owing to Theorem 2.8, choose the property $\mathfrak{D}_\rho(x)$: $\langle x|x\rangle^{-1} tr_E |x\rangle\langle x| = \rho_S$. We read it as "$\langle x|x\rangle^{-1} tr_E |x\rangle\langle x| = \rho_S$ *defines a state of the subsystem $\mathscr{S}$ of $\mathscr{C}$.*" Similarly "$\langle x|x\rangle^{-1} tr_E |x\rangle\langle x| = \rho'_S$ *defines a state of the subsystem $\mathscr{S}$ of $\mathscr{C}$*"; but the two states are different if $\rho_S \neq \rho'_S$, so that we must label the states in order to distinguish them. We can use as labels the density operators.

Therefore, we write the first sentence as "$\langle x|x\rangle^{-1} tr_E |x\rangle \langle x| = \rho_S$ *defines the state* ρ_S *of the subsystem* $\mathcal{S}$ of $\mathcal{C}$." The sentence "*the state of the subsystem* $\mathcal{S}$ of $\mathcal{C}$ is ρ_S" is true (by definition) iff the sentence "*the state* $|x\rangle$ *of* $\mathcal{C}$ *is such that* $\langle x|x\rangle^{-1} tr_E |x\rangle \langle x| = \rho_S$ *is satisfied*" is true.

We remember that the composite system is "the Universe," so that it is in some state defined by a ray of $\mathbb{P}(\mathcal{H} \otimes \mathcal{K})$, and the prescription of a state of the subsystem $\mathcal{S}$ of $\mathcal{C}$ is simply a constraint on the possible states of $\mathcal{C}$. Therefore, the complete specification of the state of the subsystem $\mathcal{S}$ of $\mathcal{C}$ can be regarded as a partial specification of the state of $\mathcal{C}$.

Chapter 3
Relation Between the State of a System as Isolated and as Open

Naively, we can consider $\mathscr{S}$ sometimes as a subsystem of $\mathscr{C}$ and sometimes as a closed system. Our next goal is to arrive at a sharp mathematical formulation of the physical conditions under which the first or the second point of view can be adopted. A necessary condition for the second possibility is the assumption that no interactions between the system and the environment take place. This sentence implies the idea of a transformation: we say that there is no interaction when the physical conditions are such that only specific types of transformations are allowed for the composite system, i.e., those which can be represented as a tensor product of a transformation of $\mathscr{U}(\mathscr{H})$ and a transformation of $\mathscr{U}(\mathscr{K})$. As already discussed, a transformation involves two states: a state "before" the transformation and a state "after" the transformation.

If we assume that $\mathscr{S}$ can be considered as isolated, there is no interaction and its states are modeled as elements of $\mathbb{P}(\mathscr{H})$. Standard quantum mechanics assumes that there is $T \in \mathscr{U}^{Z}(\mathscr{H})$ such that the possible pairs (state before, state after) are the pairs $(\zeta, T\zeta)$ with $\zeta \in \mathbb{P}(\mathscr{H})$. If we assume that $\mathscr{S}$ is not isolated, in general, it is only at the level of $\mathscr{C}$ (which is closed) that the states can be modeled as points of a projective space (in this case $\mathbb{P}(\mathscr{H} \otimes \mathscr{K})$) but not at the level of $\mathscr{S}$. In this case, the state of $\mathscr{S}$ before is a density operator ρ_S and the state of $\mathscr{S}$ after is a density operator ρ'_S.

Furthermore, if an interaction takes place, it is only at the level of $\mathscr{S}$ that the states before and after are connected by a transformation of $\mathscr{U}^{Z}(\mathscr{H} \otimes \mathscr{K})$; the density operators ρ_S and ρ'_S are not connected by any simple relation: they are just "projections" (through the partial trace operation) of the initial and final states of $\mathscr{C}$. To be more specific, if we represent the "evolution" of $\mathscr{C}$ by $T \in \mathscr{U}(\mathscr{H} \otimes \mathscr{K})$, and the state of $\mathscr{C}$ "before" with the normalized ket $|\psi\rangle$, the state of $\mathscr{C}$ "after" is

The content of this chapter can be found in Campanella's file *ZurekState* (2011-09-01).

M. Campanella et al., *Interpretative Aspects of Quantum Mechanics*, UNIPA Springer Series,
https://doi.org/10.1007/978-3-030-44207-1_3

represented by $|\psi'\rangle = T|\psi\rangle$. The states of $\mathscr{S}$ as a subsystem of $\mathscr{C}$ "before" and "after" are $\rho_S = tr|\psi\rangle\langle\psi|$, and $\rho'_S = tr|\psi'\rangle\langle\psi'|$.

Even if $|\psi\rangle = |u\rangle\,|v\rangle$, i.e., "before" both $\mathscr{S}$ and $\mathscr{E}$ are in well-defined states as isolated systems, so that $\rho_S = |u\rangle\,\langle u|$, ρ'_S, in general, is no more a one-dimensional projector. Therefore, while the state before can be represented by a point of $\mathbb{P}(\mathscr{H})$, the state "after" is, generally speaking, a mathematical object of a different nature.

3.1 Representation of Isolated and Non-isolated Systems

There are two related, but distinct items raised by this discussion. The first consists in the different ways of representing the state of $\mathscr{S}$ depending on whether it is considered isolated or not. Classically, the state of a composite system is described as a point in its phase space, which is expressed as the Cartesian product of the phase spaces of the component subsystems.

If also in the classical case we define the state of $\mathscr{S}$ *a la Zurek*, with the role of $\mathscr{U}^Z(\mathscr{K})$ substituted by the role of the group of the canonical transformations of the phase space of $\mathscr{E}$, we see that the orbits of the latter group (i.e., the states *a la Zurek* of $\mathscr{S}$ as a subsystem of $\mathscr{C}$) can be parameterized by the points of the phase space of $\mathscr{S}$. This entails that the state of the system at each time is represented by the same object (a point in its phase space) both in the absence and in the presence of interaction. So in the classical case the conventional definition of states of the subsystem is compatible with Zurek's point of view.

In the latter case, the projections of the initial and final points of the composite system are points of the phase space of $\mathscr{S}$ even if an interaction takes place. It is true that the final point of $\mathscr{S}$ depends both on the initial states of $\mathscr{S}$ and of $\mathscr{E}$, so that, if the state of the latter is not completely known (e.g., because $\mathscr{E}$ is a complex system), statistical methods, based on the introduction of the phase density, must be introduced. But the description of the state of $\mathscr{S}$ by means of a point in its phase space is not forbidden, at least *in principle*. The situation is quite different in the quantum case. Even if the initial state is described by a point of $\mathbb{P}(\mathscr{H})$, the final state of $\mathscr{S}$ (i.e., the "projection" of the final state of the composite system) is a density operator with rank > 1 and not just a one-dimensional projector (which is equivalent to a point of $\mathbb{P}(\mathscr{H})$) if an interaction with the environment has taken place.

The second question concerns the different ways of describing the evolution of the system in the absence or in the presence of interaction. In this respect, the classical and the quantum cases are similar; in both cases, the interaction precludes the possibility of establishing a functional relationship between the initial and the final states of $\mathscr{S}$: the only thing that can be said, in general, is that both are "projections" of the initial and final states of the composite system.

Hence, the main conclusion of the above discussion is that the basic difference between the classical and the quantum case in this respect is that in the latter a fundamental difference exists in the mathematical description of the states of $\mathscr{S}$ depending on whether the system is considered as isolated or as a subsystem of $\mathscr{C}$. In the first case, the states are points of $\mathbb{P}(\mathscr{H})$, while in the second they are represented by density operators. Of course, in the most popular axiomatics of quantum mechanics the need of this difference is a consequence of the a priori introduction in the axioms of a probabilistic interpretation of the theory and of its quantitative expression through Born's rule.

We emphasize that instead, in our analysis, this difference is a direct consequence of two assumptions: the first is that a composite system is described by a tensor product, and the second is that the correct notion of state of a subsystem is the one implied by Zurek's considerations (we emphasize that the "philosophy" underlying the notion of the state of a subsystem is identical in the classical and in the quantum case: it is a property of the state of the composite system which is left invariant by all the transformations that involve only the environment, such that every other invariant property is a consequence of it). Neither of these assumptions involve probability; specifically, the density operator arises merely as an *orbit invariant*. We will see in the following that we are *forced* to give it a probabilistic interpretation and that its quantitative implementation is equivalent to Born's rule. To carry on this program, we carefully investigate the relation between the two possible descriptions of the states of $\mathscr{S}$.

Suppose we have good reasons to consider the system isolated from the environment before some instant t_0, but that this assumption is no longer valid after, up to the instant t_1, while for $t > t_1$ $\mathscr{S}$ and $\mathscr{E}$ can be again modeled as mutually isolated. Before t_0, $\mathscr{S}$ and $\mathscr{E}$ evolve independently; at $t < t_0$, $\mathscr{S}$ is in a well-definite state $\varsigma \in \mathbb{P}(\mathscr{H})$ and $\mathscr{S}$ in a state $\tau \in \mathbb{P}(\mathscr{K})$. At $t' < t_0$, we have $\varsigma' = T\varsigma$ and $\tau' = S\tau$, with $T \in \mathscr{U}^Z(\mathscr{H})$ and $S \in \mathscr{U}^Z(\mathscr{K})$. If the normalized kets $|\sigma\rangle$ and $|\theta\rangle$ represent ς and τ, respectively, and $|\sigma'\rangle$ and $|\theta'\rangle$ represent ς' and τ' respectively, there are U representing T and V representing S such that $|\sigma'\rangle = U\,|\sigma\rangle$ and $|\theta'\rangle = V\,|\theta\rangle$. We read the pair $(|\sigma\rangle, |\theta\rangle)$ saying that at t $\mathscr{S}$ is in the state ς and $\mathscr{E}$ is in the state τ. At the same instant, $\mathscr{C}$ is in a state ζ represented by $|\sigma\rangle\,|\theta\rangle$. We note that $tr_E\,|\sigma\rangle\langle\sigma| \otimes |\theta\rangle\langle\theta| = |\sigma\rangle\langle\sigma|$ so that the density operator is a projector on a one-dimensional space, which is a description of a pure state equivalent to a ray.

Similarly, at t' $\mathscr{S}$ is in the state ς' and $\mathscr{E}$ is in the state τ', while $\mathscr{C}$ is in a state ζ' represented by $U \otimes V\,|\sigma\rangle\,|\theta\rangle$. At t_0 the interaction is switched on. Therefore, the state of $\mathscr{C}$ after t_0 becomes entangled. In particular, let $|\psi\rangle$ denote a normalized representative of the state at t_1. At this instant, the system must be considered as interacting with the environment as the "past" is concerned, and as non-interacting regards to the "future."

If "non-interacting" were the same thing as "isolated," we may think that both descriptions of the state must hold: as an interacting system it ought to be described by $\rho_S = tr_E\,|\psi\rangle\langle\psi|$, (which means a partial specification of the state of $\mathscr{C}$) and as a non-interacting system by a (perhaps partial) specification of some point of ξ of $\mathbb{P}(\mathscr{H})$. A careful analysis will, however, show that not for all the density operators the description in terms of states of an isolated system is possible. We will, in particular, find that, when there is in ρ_S some kind of degeneracy to be specified later, an obstruction to such a description arises, so that the notions of "non-interacting" and "isolated" are not equivalent.

3.2 Conditions for Density Operators

We now assume that ρ_S is such that a description is possible, and we find the conditions which must be satisfied.

If in the same physical situation ρ_S is a state of the system as subsystem of $\mathscr{C}$ and ξ a state of $\mathscr{S}$ as an isolated system, we expect that there is a relation between ρ_S and ξ. Namely, for each allowed ρ_S there is a set Ξ of possible values for ξ such that, if it is true that the state of $\mathscr{S}$ as a subsystem of $\mathscr{C}$ is ρ_S, the state of $\mathscr{S}$ as an isolated system is an element of the set Ξ.

Let us consider a transformation of $\mathscr{C}$ represented by $U \otimes V$ which leaves ρ_S invariant. Then there is no evolution in $\mathscr{S}$ as a subsystem of $\mathscr{C}$. The above transformation indicates that there is no interaction, and we suppose that the description of $\mathscr{S}$ as an isolated system is also valid. In this description too there must be no evolution. The transformation law of the density operator is $U\rho_S U^*$. Let $\rho_S = \sum p_\alpha P_\alpha$ be the spectral expansion of ρ_S. The transformation U leaves ρ_S invariant iff $U P_\alpha U^* = P_\alpha$, $\forall\alpha$.

If P_0 is the projector on the null space of ρ_S, then $P_0 = I - \sum P_\alpha$ and $U P_0 U^* = P_0$. On the other hand, in the isolated system description, the new state is $T\xi$, where $T \in \mathscr{U}^Z(\mathscr{H})$ is represented by U. Hence, for each possible state ξ of $\mathscr{S}$ as an isolated system given that the state of $\mathscr{S}$ as a subsystem of $\mathscr{C}$ is ρ_S, the new state $T\xi$ must be equal to ξ.

Let $\mathscr{G}_\rho$ be the stabilizer of ρ, and $\mathscr{G}_\xi$ the stabilizer of ξ. Hence, $\mathscr{G}_\rho \subseteq \mathscr{G}_\xi$ for each $\xi \in \Xi$, so that

$$\mathscr{G}_\rho \subseteq \bigcap_{\xi\in\Xi} \mathscr{G}_\xi \triangleq \Gamma_\Xi. \tag{3.2.1}$$

If $|\sigma\rangle \in \mathscr{H}$ represents ξ, and if $U\rho_S U^* = \rho_s$, then $U|\sigma\rangle = \lambda|\sigma\rangle$ ($\lambda \in \mathbb{C}$, $|\lambda| = 1$) (where λ may depend on U); furthermore, $|\sigma\rangle$ has an expansion of the form $|\sigma\rangle = \sum P_\alpha |\sigma\rangle + P_0 |\sigma\rangle$. Hence we must have

$$\sum U P_\alpha |\sigma\rangle + U P_0 |\sigma\rangle = \sum \lambda P_\alpha |\sigma\rangle + \lambda P_0 |\sigma\rangle. \tag{3.2.2}$$

Multiplying both sides by P_β we get

$$U P_\beta |\sigma\rangle = \lambda P_\beta |\sigma\rangle, \tag{3.2.3}$$

and multiplying both sides by P_0,

$$U P_0 |\sigma\rangle = \lambda P_0 |\sigma\rangle. \tag{3.2.4}$$

But the restrictions of U to the images of the projectors are arbitrary unitary transformations of these images, so that if $P_\beta |\sigma\rangle$ is nonzero, Im P_β is one-dimensional. The same conclusion holds for $P_0 |\sigma\rangle$.

Furthermore, if $P|\sigma\rangle \neq 0$ and $P'|\sigma\rangle \neq 0$ with $P \neq P'$, we can choose U such that $UP|\sigma\rangle = e^{i\varphi}P|\sigma\rangle$ and $UP'|\sigma\rangle = e^{i\varphi'}P'|\sigma\rangle$ with $e^{i\varphi} \neq e^{i\varphi'}$. But this entails that $e^{i\varphi} = \lambda$ and $e^{i\varphi'} = \lambda$ and this is a contradiction. We conclude that, for each possible state ξ of $\mathscr{S}$ as an isolated system, given that the state of $\mathscr{S}$ as a subsystem of $\mathscr{C}$ is ρ_S, ξ is a one-dimensional eigenspace of ρ_S. Then Ξ is a set of eigenstates of ρ_S.

The subgroup of $\mathscr{U}^Z(\mathscr{H})$ which leaves invariant all the elements of Ξ is nothing but Γ_Ξ. No evolution is observed in $\mathscr{S}$ under Γ_Ξ if the only possible states of it as an isolated system are the elements of Ξ. Therefore, no evolution must be observed in $\mathscr{S}$ as a subsystem of $\mathscr{C}$ in the state ρ_S. Consequently, $\Gamma_\Xi \subseteq \mathscr{G}_\rho$, so that

$$\Gamma_\Xi = \mathscr{G}_\rho. \tag{3.2.5}$$

Hence, we must determine the pairs (ρ, Ξ) satisfying the above condition. If the orthogonal complement $\mathscr{L}_{\Xi\perp}$ of the space $\mathscr{L}_\Xi$ generated by Ξ is nonzero, the restriction of Γ_Ξ to it acts on it as the full unitary group. But $\mathscr{L}_{\Xi\perp}$ is an invariant subspace of ρ_S; as the restriction to it of $\mathscr{G}_\rho$ is the full unitary group, it must consist of a single eigenspace. We conclude that, besides the elements of Ξ, there is at most a single eigenspace. If $\mathscr{L}_{\Xi\perp} = 0$, the set of eigenspaces of ρ_S is Ξ, so that they all are one-dimensional; if the latter is at least two-dimensional, the set of the one-dimensional eigenspaces of ρ_S is Ξ. Finally, if $\mathscr{L}_{\Xi\perp} = 0$ is one-dimensional, the eigenspaces of ρ_S consist of Ξ and in addition the orthogonal complement of $\mathscr{L}_{\Xi\perp}$. All the above possibilities can be summarized in the following.

Theorem 3.1 *For each orthogonal set Ξ of rays, the density operators satisfying the equation $\Gamma_\Xi = \mathscr{G}_\rho$ are all those whose eigenspaces are the elements of Ξ and the orthogonal complement of the subspace generated by them.*

Remark 3.1 *Remark on terminology:* In order to avoid confusion, we remark that we call *invariant* space of an operator a subspace which is transformed into itself by the operator, while we call *eigenspace* of the operator a *maximal* invariant space. As it is usual, we call *eigenvector* of the operator a nonzero vector on which the operator acts as a scalar multiplier. According to our terminology, the one-dimensional subspace generated by an eigenvector is an eigenspace only if the associated eigenvalue is simple, otherwise it is only an invariant space. An *eigenstate* is a one-dimensional eigenspace. (Using this terminology, the null space of a density operator is the eigenspace corresponding to the zero eigenvalue.)

We can write the most general expression of a density operator consistent with the conditions imposed as yet. This expression is an immediate consequence of Theorem 2.8. We have

$$\rho_S = \sum_{\xi\in\Xi} w(\xi)\,|\xi\rangle\,\langle\xi| + gP, \tag{3.2.6}$$

where $I = \sum_{\xi \in \Xi} |\xi\rangle\langle\xi| + P$, $\Xi \xrightarrow{w} \mathbb{R}_{\geq}$ is injective and $g \notin \mathrm{Im}\, w$ with $g \geq 0$, and with the normalization condition $\sum_{\xi \in \Xi} w(\xi) + g = 1$.

We note that, given Ξ, the set of eigenspaces of ρ_S is completely specified. It is given by the union of Ξ and the orthogonal complement of the space generated by it. The possible ρ_S are then specified by prescribing at will the set of nonnegative different eigenvalues, satisfying the normalization condition.

The above equation shows that the orthogonal set of rays can be chosen at will and expresses the general solution for ρ_S. Conversely, suppose that ρ_S is prescribed and that we want to find all the sets Ξ satisfying the equation $\Gamma_\Xi = \mathcal{G}_\rho$. By Theorem 2.8, the eigendecomposition of ρ_S includes at most one eigenspace at least two-dimensional. If such a space appears in the eigendecomposition, Ξ is given by the set of the remaining eigenspaces; if all the eigenspaces are one-dimensional, Ξ is the whole set of eigenspaces with the possible exception of one of them. We can summarize the situation as follows.

Theorem 3.2 *When the system $\mathcal{H}$ in ρ_H can be considered isolated, there is in ρ_H at most a single eigenspace at least two-dimensional; the set Ξ of possible states of $\mathcal{H}$ as an isolated system is the set of eigenstates of ρ_H if such an eigenspace exists; if all the eigenspaces of ρ_H are one-dimensional, the set Ξ consists of all these eigenspaces with the possible exception of one of them.*

A further analysis that we will soon present will remove this indeterminacy.

We remember that $\mathcal{H}$ is a subsystem of $\mathcal{C}$, and that the latter is the model of our "Universe," so that its state is represented by some $|\psi\rangle \in \mathcal{C}$ such that $\rho_H = tr_E |\psi\rangle\langle\psi|$. We now investigate how the conditions just found for ρ_H are reflected in the canonical decomposition of $|\psi\rangle$. The latter can be put in the form

$$|\psi\rangle = \sum_{\chi \in X} d(\chi)\, |\chi\rangle, \tag{3.2.7}$$

with $tr_E |\chi\rangle\langle\chi'| = 0$ for $\chi \neq \chi'$ and $tr_E |\chi\rangle\langle\chi| = P(\chi)$. We get

$$\rho_H = \sum_{\chi \in X} d^2(\chi)\, P(\chi). \tag{3.2.8}$$

The projectors appearing in the above expression correspond to the non-null eigenspaces of ρ_H; they must be one-dimensional but at most one of them, and hence the kets $|\chi\rangle$ are indecomposable, but at most one. Each indecomposable $|\chi'\rangle$ has the form $|\xi'(\chi')\rangle\, |\eta'(\chi')\rangle$, so that we get

$$|\psi\rangle = \sum_{\chi' \in X'} d(\chi')\, |\xi'(\chi')\rangle\, |\eta'(\chi')\rangle + d\, |\delta\rangle, \tag{3.2.9}$$

where $d = 0$ if all the kets are indecomposable. We have $P(\chi') = |\xi'(\chi')\rangle\langle\xi'(\chi')|$, so that $|\xi'(\chi')\rangle$ and $|\xi'(\chi'')\rangle$ are orthogonal and hence different for $\chi' \neq \chi''$. The same conditions hold for the $|\eta\rangle$.

Imagine that $\mathscr{E}$ is in turn a composite system, say $\mathscr{E} = \mathscr{K} \otimes \mathscr{L}$. We now introduce a special type of state in $\mathscr{H} \otimes \mathscr{K} \otimes \mathscr{L}$.

Definition 3.1 We say that a ket $|\phi\rangle \in \mathscr{H} \otimes \mathscr{K} \otimes \mathscr{L}$ represents an M-state if there is a set A, a partition $\mathscr{P}_2$ of A, three injections $\mathscr{P}_2 \xrightarrow{\kappa} \mathscr{H}$, $A \xrightarrow{\eta} \mathscr{K}$, $A \xrightarrow{\zeta} \mathscr{L}$ satisfying the conditions $\langle\kappa(y)|\kappa(y')\rangle = \delta_{yy'}$, $\langle\eta(a)|\eta(a')\rangle = \delta_{aa'}$, $\langle\zeta(a)|\zeta(a')\rangle = \delta_{aa'}$, and an injective mapping $A \xrightarrow{\upsilon} \mathbb{R}_>$ such that

$$|\phi\rangle = \sum_{a\in A} \upsilon(a)\,|\kappa\circ\pi_2(a)\,\eta(a)\,\zeta(a)\rangle, \tag{3.2.10}$$

where π_2 is the canonical projection associated with the partition $\mathscr{P}_2$.

The M-state obtained in the way described above will be denoted $(A, \mathscr{P}_2, \kappa, \eta, \zeta, \upsilon)$. We prove the following.

Theorem 3.3 *$(A, \mathscr{P}_2, \kappa, \eta, \zeta, \upsilon)$ and $(A', \mathscr{P}'_2, \kappa', \eta', \zeta', \upsilon')$ define the same M-state if and only if there is a bijection $A \xrightarrow{j} A'$ such that $\upsilon = \upsilon' \circ j$, $\mathscr{P}'_2 = j\mathscr{P}_2$, and mappings $\mathscr{P}_2 \xrightarrow{\lambda} \mathbb{C}_1$, $A \xrightarrow{\mu} \mathbb{C}_1$, $A \xrightarrow{\nu} \mathbb{C}_1$ with $(\lambda\circ\pi_2)\mu\nu = 1$ such that $\kappa\circ\pi_2 = (\lambda\circ\pi_2)\kappa'\circ\pi'_2\circ j$, $\eta = \mu\eta'\circ j$ and $\zeta = \nu\zeta'$. The mappings j, λ, μ, ν are unique.*

Proof The expansion that defines $|\phi\rangle$ is its canonical expansion when $|\phi\rangle$ is regarded as an element of $(\mathscr{H} \otimes \mathscr{K}) \otimes \mathscr{L}$. Hence, the images of υ and υ' are the same. Consequently, the equation $\upsilon = \upsilon' \circ j$ has $\upsilon'^{-1} \circ \upsilon$ as its unique solution. Furthermore, the kets appearing in the expansion are in bijection with the coefficients, so that $|(\kappa\circ\pi_2)\,\eta\zeta\rangle = |[(\kappa'\circ\pi'_2)\,\eta'\zeta']\circ j\rangle$. The single factors in the tensor product are defined up to phase factors whose product is 1, and hence the thesis. □

We have

$$\rho_{HK} = \sum_{a\in A} \upsilon(a)^2\,|\kappa\circ\pi_2(a)\,\eta(a)\rangle\,\langle\kappa\circ\pi_2(a)\,\eta(a)|.$$

We observe that the dimension of the space generated by all the

$$|\theta(a)\rangle \triangleq |\kappa\circ\pi_2(a)\,\eta(a)\rangle$$

is equal to $|A|$.

The dimension of $\mathscr{K}$ is at least $|A|$, and at least 2, i.e., $\dim\mathscr{K} \geq \max(2, |A|)$, while the dimension of $\mathscr{H}$ is at least 2. Hence, the dimension of $\mathscr{H} \otimes \mathscr{K}$ is at least $2\max(2, |A|)$ and the codimension of the space generated by all the $|\theta(a)\rangle$ is at least $2\max(2, |A|) - |A|$. Consequently, this codimension is at least $|A|$ if $|A| \geq 2$ and at least 3 if $|A| = 1$. In any case, the codimension is at least 2. Remembering

Theorem 3.2, we conclude in every case that the set of possible states associated with ρ_{HK} is its set of eigenstates, which in this case are nothing but those appearing in the spectral decomposition of ρ_{HK}, i.e., non-null eigenstates.

Let us now evaluate ρ_H. We have

$$\rho_H = tr_K \rho_{HK} = \sum_{a \in A} v(a)^2 \left|\kappa \circ \pi_2(a)\right\rangle \left\langle \kappa \circ \pi_2(a)\right| = \sum_{y \in P_2} \left|\kappa(y)\right\rangle \left\langle \kappa(y)\right| \sum_{a \in y} v(a)^2. \tag{3.2.11}$$

Let $\widetilde{w}(y) \triangleq \sum_{a \in y} v(a)^2$. If $\mathscr{P}_1$ is the partition induced on $\mathscr{P}_2$ by $\widetilde{w}$, we can uniquely define an injective mapping $\mathscr{P}_1 \xrightarrow{w} \mathbb{R}_>$ such that $\widetilde{w}(y) = w(x)$ whenever $y \in x$. Hence we have

$$\rho_H = \sum_{y \in P_2} \widetilde{w}(y) \left|\kappa(y)\right\rangle \left\langle \kappa(y)\right| = \sum_{x \in P_1} \sum_{y \in x} w(x) \left|\kappa(y)\right\rangle \left\langle \kappa(y)\right|. \tag{3.2.12}$$

Putting $P(x) \triangleq \sum_{y \in x} \left|\kappa(y)\right\rangle \left\langle \kappa(y)\right|$ we get

$$\rho_H = \sum_{x \in P_1} w(x) P(x), \tag{3.2.13}$$

which, remembering the injectivity of w, is recognized to be the spectral representation of ρ_H. We will now show the theorem as follows.

Theorem 3.4 *If ρ_H is an arbitrary density operator of $\mathscr{H}$, there are Hilbert spaces $\mathscr{K}$ and $\mathscr{L}$ together with an M-state $|\phi\rangle \in \mathscr{H} \otimes \mathscr{K} \otimes \mathscr{L}$ such that $\rho_H = tr_{KL} |\psi\rangle \langle\psi|$.*

Indeed, suppose that $\rho_H = \sum_{w \in W} w P(w)$ is the spectral representation of ρ_H. For each w, we choose an orthonormal basis $\widetilde{Y}_w$ for the corresponding eigenspace, so that $P(w) = \sum_{\widetilde{y} \in \widetilde{Y}_w} |\widetilde{y}\rangle \langle\widetilde{y}|$. We define $\widetilde{Y} = \bigcup_{w \in W} \widetilde{Y}_w$ and $\widetilde{Y} \xrightarrow{\widetilde{w}} \mathbb{R}_>$ as $\widetilde{w}(\widetilde{y}) = w$ for $\widetilde{y} \in \widetilde{Y}_w$.

Let us further associate to each $\widetilde{y}$ a set $y \triangleq A(\widetilde{y})$, with the condition that these sets are mutually disjoint, and put $A = \bigcup_{\widetilde{y} \in \widetilde{Y}} A\left(\widetilde{Y}\right)$. There is a bijection between the y and the $\widetilde{y}$, so that we can put $\widetilde{y} = \kappa(y)$. We denote $\mathscr{P}_2$ the partition of A in the sets y. We introduce an injective mapping $A \xrightarrow{\gamma} \mathbb{R}_>$ satisfying the condition $\sum_{a \in y} \gamma(a) = \widetilde{w}(\kappa(y))$. Such a mapping exists if and only if the cardinality of y is at least 2 for each y associated with a multiple eigenvalue of ρ_H, so we choose the sets according to this criterion.

We now choose two spaces $\mathscr{K}$ and $\mathscr{L}$ whose dimensionality is at least $|A|$ and two injections $A \xrightarrow{\eta} \mathscr{K}$ and $A \xrightarrow{\zeta} \mathscr{L}$ such that their images are orthonormal sets. We put $\upsilon(a) = \sqrt{\gamma(a)}$ and define

$$|\phi\rangle = \sum_{a \in A} \upsilon(a) \, |\kappa \circ \pi_2(a) \, \eta(a) \, \zeta(a)\rangle. \qquad (3.2.14)$$

A straightforward computation shows that $\rho_H = tr_{KL} \, |\phi\rangle \langle\phi|$. □

Suppose now that $\mathscr{H}$ in ρ_H can be considered isolated. The state of $\mathscr{H}$ as a subsystem of a larger system is completely described by ρ_H, so that the possibility of considering it as isolated and the set of its possible states depend uniquely on ρ_H and not on the specific realization of ρ_H in terms of a larger system. Theorem 3.4 shows that $\rho_H = tr_{KL} \, |\phi\rangle \langle\phi|$ for some M-state $|\phi\rangle$.

We can show that, if among the eigenspaces of ρ_H there is a (single) multidimensional eigenspace, the corresponding eigenvalue must be zero. Indeed, suppose the contrary. The corresponding projector is $P(x) \triangleq \sum_{y \in x} |\kappa(y)\rangle \langle\kappa(y)|$ for some x. Hence, the eigenspace contains vectors $|\kappa(y)\rangle$ appearing as factors in the expansion of ρ_{HK}. As each term of the decomposition is a possible state of $\mathscr{H} \otimes \mathscr{K}$ as an isolated system and it is indecomposable, the vectors appearing in $P(x)$ belong to Ξ. Thus, we have found states of Ξ belonging to a multidimensional eigenspace of ρ_H and this is a contradiction. We conclude that, if all the eigenspaces of ρ_H are one-dimensional, the possible states of $\mathscr{H}$ are exactly the non-null eigenspaces.

The results of our discussion imply the following theorem.

Theorem 3.5 *If the system $\mathscr{H}$ can be considered isolated in ρ_H, then the only multidimensional eigenspace of ρ_H, if it exists, is the null eigenspace. The spectral expansion of ρ_H is $\rho_H = \sum wP(w)$ where all the projectors $P(w)$ are of rank one, and the possible states of $\mathscr{H}$ in ρ_H as an isolated system are the states associated with these projectors.*

We now introduce some terminology. A density operator whose non-null eigenspaces are all one-dimensional will be called a *generic density operator*.

The one-dimensional eigenspaces of ρ_S are the *eigenstates* of ρ_S. A generic density operator always possesses eigenstates. They are the eigenspaces associated with the nonzero eigenvalues and in addition the null space if it is one-dimensional. Furthermore, it possesses at most one degenerate eigenvalue (the zero eigenvalue). Therefore, the set Ξ is nonempty for a generic density operator: it consists of its non-null eigenstates.

The number of eigenstates of a generic density operator will be called its *type*. The dimension of its image is its *rank*. Let m be the type of a generic density operator, r its rank, and n the dimension of the Hilbert space on which it operates. If $\rho_S = \sum p_\alpha P_\alpha$ is the spectral decomposition of ρ_S, all the projectors appearing in this expansion are eigenstates, so that the rank is the number of terms of the expansion, i.e., the number of nonzero eigenvalues, or else the number of non-null eigenstates. Hence $m \geq r$. Of course $m \leq r + 1$ and if $m = r + 1$, $m = n$ because otherwise the multiplicity of the zero eigenvalue would be greater than one. Thus, if $m = r$, $n \geq m$, while, if $m = r + 1$, $n = m$. Equivalently, $n \geq r$ and if $n = r + 1$, $m = r + 1$, while if $n \neq r + 1$, $m = r$. Else, $n \geq m$ and if $n = m$, either $r = n$ or $r = n - 1$, while if $n > m, r = m$.

3.3 Physical Interpretation of Generic Density Operators

In what follows, our goal is to obtain a *complete physical interpretation of a density operator when it is generic*. The interpretation for an arbitrary density operator will be obtained successively.

Theorem 3.5 states that $\mathscr{H}$ can be considered isolated in ρ_H only if the latter is a generic density operator; in this case, the set Ξ of its possible states as an isolated system is the set of non-null eigenstates of ρ_H, but none of them, in particular, is specified.

The situation is similar to the case when a number is constrained, say, by a second degree equation: a set of possibilities is selected (the set of the solutions), but not any, in particular, is specified.

Hence, we are *forced* to consider the state of $\mathscr{H}$ in ρ_H as an isolated system as a *variable* with values in Ξ. We have seen that the set Ξ is uniquely specified by ρ_H. When the set of its possible values is finite, we can regard a variable as an equivalence class of random variables, calling equivalent any two random variables with the same domain of their probability laws. We further note that ρ_H is specified not only by Ξ, but also by the set of eigenvalues. We will show in what follows that this set is enough to fix uniquely a probability law on Ξ, so that the state is a well-defined *random variable*. Namely, the result of our analysis will be that of the conventional interpretation of the density operator: we will find that the probability of each state is equal to the corresponding eigenvalue of the density operator. We emphasize that this result *will be derived* without any additional axiom such as, for instance, a Born's rule. On the contrary, *Born's rule will be recovered as a consequence*.

We can synthesize the conceptual framework emerging from the above discussion as follows. The composite system $\mathscr{C}$, as long as it is considered a "Universe," is in a state represented as a normalized ket $|\psi\rangle$ which is defined up to an arbitrary phase factor. We call this state generic if the corresponding density operator ρ_H is generic. Suppose that the state of $\mathscr{C}$ is generic. Hence, its canonical expansion has the form

$$|\psi\rangle = \sum d\, |\xi\,(d)\, \eta\,(d)\rangle. \tag{3.3.1}$$

Unless $|\psi\rangle$ is indecomposable, neither the state of $\mathscr{S}$ nor the state of $\mathscr{E}$ as states of isolated systems can be specified. Instead, the state of $\mathscr{S}$ as a subsystem of $\mathscr{C}$ in the sense of Zurek can be specified. This state can be characterized by the density operator

$$\rho_H = tr_K\, |\psi\rangle\, \langle\psi| = \sum d^2\, |\xi\,(d)\rangle\, \langle\xi\,(d)|. \tag{3.3.2}$$

Nevertheless, as long as no interactions take place between $\mathscr{S}$ and $\mathscr{C}$, the two subsystems can be considered as isolated, so that each of them possesses a state as an isolated system. But such a state cannot be specified, so that it must be considered as a variable. The range of this variable can be read off on the canonical decomposition. Namely, the range of the state of $\mathscr{S}$ is the set $\{|\xi\ (d)\rangle\}$, while the range of the state of $\mathscr{E}$ is the set $\{|\eta\ (d)\rangle\}$. We have already anticipated that a unique law of probability is associated to these variables, so that they will become random variables.

Chapter 4
The Probability Law for Generic Density Operators

In this last chapter, our effort will be dedicated to the determination of the probability law of the non-null eigenstates of a generic density operator, by showing that, given the composite system and its subsystem $\mathscr{H}$, a mapping f arises which associates to each generic density operator ρ_S the probability distribution of its non-null eigenstates. To afford this problem, it is useful to introduce some preliminary notions.

4.1 Mathematical Structure of a "Generic" Density Operator

Let us consider an orthogonal set Ξ of states of $\mathscr{H}$ as an isolated system. Let D_Ξ be the set of generic density operators having Ξ as the set of their non-null eigenstates. All the elements of D_Ξ have the same rank $m = |\Xi|$. If P_σ is the projector associated with the state $\sigma \in \Xi$, any $\rho_S \in D_\Xi$ has the spectral expansion

$$\rho_S = \sum_{\sigma \in \Xi} p_\sigma P_\sigma, \tag{4.1.1}$$

and through this equation D_Ξ is in bijection with the set of the injective mappings $\mathbf{p}$ from Ξ to the set $\mathbb{R}_>$ of positive real numbers such that $\sum_{\sigma \in \Xi} \mathbf{p}(\sigma) = 1$.

The set D_Ξ is a proper subset of the set T_Ξ of all the density operators which can be expressed in the form (4.1.1), with the coefficients p_σ belonging to $\mathbb{R}_\geq$, and relaxing the injectivity assumption on the mappings $\mathbf{p}$. The barycentric coordinates p_σ introduce in T_Ξ the structure of an $(m-1)$-simplex together with the standard topology. With respect to the standard topology, $\bar{D}_\Xi = T_\Xi$.

The content of this chapter can be found in Campanella's file *ZurekState* (2011-09-01).

M. Campanella et al., *Interpretative Aspects of Quantum Mechanics*, UNIPA Springer Series,
https://doi.org/10.1007/978-3-030-44207-1_4

It is useful to describe in detail the structure of a set D_Ξ which comes out from some theorems.

Theorem 4.1 *If the points $\rho \in D_\Xi$ and $\rho' \in D_\Xi$ have $\mathbf{p}$ and $\mathbf{p}'$ as the vectors of their barycentric coordinates, $\widehat{\mathbf{p}}$ and $\widehat{\mathbf{p}}'$ are the restrictions of $\mathbf{p}$ and $\mathbf{p}'$ to their images, the two points belong to the same connected component of D_Ξ if and only if $\widehat{\mathbf{p}}' \circ \widehat{\mathbf{p}}^{-1}$ is an order isomorphism with respect to the orders induced on* Im $\mathbf{p}$ *and on* Im $\mathbf{p}'$ *by the standard order in* $\mathbb{R}$.

Proof Let $\mathbf{p}$ and $\mathbf{p}'$ be such that $\widehat{\mathbf{p}}' \circ \widehat{\mathbf{p}}^{-1}$ is an order isomorphism. As the order induced in Im $\mathbf{p}$ is total, and $\widehat{\mathbf{p}}$ is bijective, it induces a total order in Ξ. As $\widehat{\mathbf{p}}' \circ \widehat{\mathbf{p}}^{-1}$ is an order isomorphism, $\widehat{\mathbf{p}}'$ induces on Ξ the same order. Under this order,

$$\sigma < \sigma' \Rightarrow p_\sigma < p_{\sigma'} \text{ and } \sigma < \sigma' \Rightarrow p'_\sigma < p'_{\sigma'}.$$

Let λ and μ be real numbers such that $\lambda \geq 0$, $\mu \geq 0$, $\lambda + \mu = 1$. If $\rho'' = \lambda\rho + \mu\rho'$, the corresponding vector of barycentric coordinates is

$$\mathbf{p}'' = \lambda\mathbf{p} + \mu\mathbf{p}'.$$

Hence

$$p''_\sigma - p''_{\sigma'} = \lambda(p_\sigma - p_{\sigma'}) + \mu(p'_\sigma - p'_{\sigma'}).$$

As λ and μ cannot be both zero, if, for instance, $\lambda \neq 0$, the condition $p''_\sigma = p''_{\sigma'}$ entails $\mu/\lambda = (p_\sigma - p_{\sigma'})/(p'_\sigma - p'_{\sigma'})$. But the numerator and the denominator of this fraction have the same sign, so we obtain an absurd. This means that the components of $\mathbf{p}''$ must be all different, so that the segment between ρ and ρ' belongs to D_Ξ and the points ρ, ρ' belong to the same arcwise connected component.

Conversely, suppose that $\widehat{\mathbf{p}}' \circ \widehat{\mathbf{p}}^{-1}$ is not an order isomorphism. This means that the orders induced by $\widehat{\mathbf{p}}'$ and $\widehat{\mathbf{p}}$ on Ξ are different, so that there are σ and σ' such that $\sigma < \sigma'$ and $\sigma <' \sigma'$. Correspondently, $p_\sigma < p_{\sigma'}$ and $p'_\sigma > p'_{\sigma'}$. Suppose that there is an arc $[0, 1] \overset{\alpha}{\to} D_\Xi$ connecting ρ and ρ'. If $\widehat{\alpha}$ is the corresponding arc in $\mathbb{R}^n$, we have $\widehat{\alpha}(0) = \mathbf{p}$ and $\widehat{\alpha}(1) = \mathbf{p}'$. Hence we have

$$\widehat{\alpha}_\sigma(0) = p_\sigma, \;\; \widehat{\alpha}_{\sigma'}(0) = p_{\sigma'}, \;\; \widehat{\alpha}_\sigma(1) = p'_\sigma, \;\; \widehat{\alpha}_{\sigma'}(1) = p'_{\sigma'}.$$

Consider $\beta(t) =: \widehat{\alpha}_\sigma(t) - \widehat{\alpha}_{\sigma'}(t)$. Then $\beta(0) = p_\sigma - p_{\sigma'} < 0$ and $\beta(1) = p'_\sigma - p'_{\sigma'} > 0$. We conclude that there is t^* such that $\widehat{\alpha}_\sigma(t^*) = \widehat{\alpha}_{\sigma'}(t^*)$ and this is in contradiction with the assumption that the arc belongs to D_Ξ. We conclude that ρ and ρ' belong to different arcwise connected components. As D_Ξ is a manifold, connectedness and arcwise connectedness coincide, so that the theorem is demonstrated.

Remark 4.1 A consequence of Theorem 4.1 is that each connected component of D_Ξ is convex. Indeed, two points belong to the same connected component iff the corresponding mapping $\widehat{\mathbf{p}}' \circ \widehat{\mathbf{p}}^{-1}$ is an order isomorphism and, if this happens, we see from the proof of Theorem 4.1 that there is in D_Ξ a segment connecting them. The points of D_Ξ can be classified according to their belonging to its connected components; they can also be classified according to the total order relations they induce in Ξ.

Theorem 4.1 establishes an injective mapping from the set of connected components of D_Ξ to the set of total order relations in Ξ. Really, this mapping is bijective. Indeed, given a total order in Ξ, we can always choose a set of values p_σ such that $\sigma < \sigma' \Rightarrow p_\sigma < p_{\sigma'}$. Hence, the connected components of D_Ξ are in bijection with the total order relations in Ξ. If $<$ is such a relation and π is a permutation of Ξ, the total order relation $<'$ defined as $\sigma <' \sigma'$ iff $\pi^{-1}\sigma < \pi^{-1}\sigma'$ will be denoted $\pi_<$.

It is easily shown that the action of the permutation group of Ξ on the total order relations is simply transitive, so that, if we fix a total order relation in Ξ, the set of all total order relations can be parameterized by the elements of the permutation group of Ξ. In particular, this shows that D_Ξ possesses $m!$ connected components.

Each permutation π defines a bijective mapping of T_Ξ into itself, and the image of its restriction to D_Ξ is D_Ξ, so that a bijective mapping $\mathfrak{m}(\pi)$ of D_Ξ into itself is obtained. The restriction of this mapping to a connected component sends it in a connected component and defines a mapping which is an isomorphism of convex spaces. Therefore, the connected components of D_Ξ are isomorphic convex spaces.

Let Δ be a connected component. As it is convex, its closure $\bar{\Delta}$ is a convex closed subspace of T_Ξ. We want to determine the structure of $\bar{\Delta}$ and its relation with Δ. As Δ is associated with a well-defined total order in Ξ, we get a well-defined isomorphism of totally ordered sets between Ξ and the set $\mathbb{N}_m$ of the first m natural numbers with the natural total order. Hence, there is a bijection which associates each point of Δ with a mapping $\mathbb{N}_m \xrightarrow{\varpi} \mathbb{R}$ such that $p_i > 0$, $\sum p_i = 1$ and $p_i < p_j$ for $i < j$. This bijection is obviously the restriction to Δ of a bicontinuous bijection defined by the same rule between T_Ξ and the standard $(m-1)$-simplex. Calling Δ_S the image of Δ, the image of $\bar{\Delta}$ is the closure $\bar{\Delta}_S$ of Δ_S and hence we have the following lemma.

Lemma 4.1 *The closure of Δ_S is the set characterized by $p_i \geq 0$, $\sum p_i = 1$, and $p_i \leq p_j$ for $i < j$.*

Indeed, let $\mathbf{p} \in \overline{\Delta_S}$. Hence, there is a sequence $\mathbf{p}_k$ of elements of Δ_S such that

$$\lim_{k\to\infty} \mathbf{p}_k = \mathbf{p}.$$

Hence, for $i < j$,

$$\lim_{k\to\infty} (p_i^{(k)} - p_j^{(k)}) = p_i - p_j,$$

so that $p_i \leq p_j$. The other conditions are obvious.

Vice versa, suppose that $\mathbf{p}$ satisfies all the conditions of the lemma. Let ε_k be a sequence with positive terms converging to zero. Define

$$p'^{(k)}_1 = p_1, p'^{(k)}_i = p_i + (i-1)\varepsilon_k \text{ for } 2 \leq i \leq m-1 \text{ and } p'^{(k)}_m = 1 - \sum_{i=1}^{m-1} p'^{(k)}_i.$$

But

$$\sum_{i=1}^{m-1} p'^{(k)}_i = 1 - p_m + \varepsilon_k \sum_{i=1}^{m-1}(i-1) = 1 - p_m + (m-1)(m-2)\varepsilon_k/2.$$

In order to ensure that $p'^{(k)}_m > 0$, we require that $\varepsilon_k < 2p_m/(m-1)(m-2)$. In this way, $p'^{(k)}_i > 0$ and $\sum p'^{(k)}_i = 1$. Furthermore,

$$p'^{(k)}_1 - p'^{(k)}_2 = p_1 - p_2 - \varepsilon_k < 0,$$
$$p'^{(k)}_i - p'^{(k)}_{i+1} = p_i - p_{i+1} - \varepsilon_k < 0 \text{ for } 2 \leq i \leq m-1.$$

The corresponding $\mathbf{p}^{(k)}$ belongs to Δ_S and $\lim_{k\to\infty} \mathbf{p}_k = \mathbf{p}$. Hence $\mathbf{p} \in \overline{\Delta}_S$.

The structures of $\bar{\Delta}_S$ and Δ_S are completely characterized by the following theorem.

Theorem 4.2 *The transformation:*

$$x'_1 = mx_1, \ \ldots., \ x'_i = (m-i+1)\,(x_i - x_{i-1})\,, \ \ldots., \ x'_m = x_m - x_{m-1}$$

induces an isomorphism of convex spaces between $\bar{\Delta}_S$ and the standard $(m-1)$-simplex. In this isomorphism Δ_S and the interior of the simplex correspond each other.

Proof We first observe that the set of conditions $x_i \leq x_j$ for $i < j$ $(i, j \in \mathbb{N}_m)$ is equivalent to the set of conditions $x_i \leq x_{i+1}$ for $i \in \mathbb{N}_{m-1}$. Indeed the second set of conditions is entailed by the first: it is sufficient to observe that we obtain the second set from the first putting $j = i+1$. Conversely observe that for $i < j$, we can write $x_j = x_i + \sum_{k=1}^{j-1}(x_{k+1} - x_k)$, from which we see immediately that the first set of conditions is a consequence of the second. Therefore, if $\mathbf{x} \in \bar{\Delta}_S$, $x'_i \geq 0$ for $1 \leq i \leq m$.

Furthermore

$$\sum_{i=1}^{m} x'_i = \sum_{i=2}^{m}(m-i+1)(x_i - x_{i-1}) + mx_1.$$

In this expression, the coefficient of x_1 is $m-(m-1) = 1$, while, for $m \geq 2$, the coefficient of x_1 is $(m-i+1)-(m-1) = 1$, so that $\sum_{i=1}^{m} x_i = \sum_{i=1}^{m} x'_i$.

Hence, $\mathbf{x}'$ belongs to the standard $(m-1)$-simplex. The transformation is composed of the transformation

$$y_1 = x_1, \quad y_i = x_i - x_{i-1} \qquad (2 \le i \le m)$$

and of the transformation

$$x'_i = (m - i + 1)y_i.$$

Both are invertible, so their composition is invertible. The inverse of the first is

$$x_1 = y_1, \quad x_i = y_i + \sum_{j=2}^{i} y_j \qquad (2 \le i \le m).$$

The inverse of the second is

$$y_i = x'_i/(m - i + 1).$$

We have therefore

$$x_1 = x'_1/m, \quad x_i = \sum_{j=1}^{i} x'_j/(m - j + 1) \qquad (2 \le i \le m).$$

As the hyperplane $\sum_{i=1}^{m} x_i = 1$ is invariant for the direct transformation, it is invariant also for the inverse transformation, so that, if $\sum_{i=1}^{m} x'_i = 1$, $\sum_{i=1}^{m} x_i = 1$.

Furthermore, if $x'_i \ge 0$ for all i, $x_i \ge 0$ for all i. Finally

$$x_i - x_{i-1} = x'_i/(m - i + 1) \ge 0 \qquad (2 \le i \le m).$$

We conclude that, if $\mathbf{x}'$ belongs to the standard $(m-1)$-simplex, $\mathbf{x} \in \bar{\Delta}_S$. Hence, the transformation induces a bijection between the two which, being the restriction of a linear transformation, is an isomorphism of convex spaces. □

The structure of Δ_S is completely clarified by the following.

Lemma 4.2 *In the transformation of Theorem 4.2 the points with the coordinates all different to each other are in bijection with the points of the standard $(m-1)$-simplex whose coordinates are all positive with the exception at most of the first.*

Proof If the x_i are all different, each

$$x'_i = (m - i + 1)(x_i - x_{i-1}) \qquad (2 \le i \le m)$$

is strictly positive. Conversely, if each x'_i $(2 \leq i \leq m)$ is strictly positive, all the inequalities $x_{i-1} \leq x_i$ are strict, so that all the x_i are different to each other.

From the above lemma, we easily prove the following theorem.

Theorem 4.3 *Δ_S is the inverse image of the interior of the standard $(m-1)$-simplex for $m < n$ and of the union of it, and the interior of the face opposite to the vertex whose coordinates are $x'_1 = 1$, $x'_i = 0$ $(2 \leq i \leq m)$ for $m = n$.*

Proof By Lemma 4.2 the image of Δ_S is the subset of the standard $(m-1)$-simplex characterized by $x'_i > 0$ $(1 \leq i \leq m)$ for $m < n$, which is the interior of the simplex. If $m = n$, the zero value of the first coordinate is allowed. If we remember that a face of the standard simplex is a set of points with a coordinate equal to zero and that the face opposite to a given vertex is the unique face that does not contain the vertex, we see that we must add to the interior of the simplex the interior of the face opposite to the vertex whose coordinates are $x'_1 = 1$, $x'_i = 0$ $(2 \leq i \leq m)$. □

Using the inverse transformation, we can find the vertices of $\overline{\Delta}_S$, which are in a one-to-one correspondence with the vertices of the standard simplex. Let Q'_i be the vertex of the latter with coordinates $x'_{ik} = \delta_{ik}$. For $i = 1$, we get from the inverse transformation found in the proof of Theorem 4.3 $x_j^{(1)} = 1/m$, while for $i > 1$ $x_j^{(i)} = 0$ for $j < i$ and $x_j^{(i)} = 1/(m-i+1)$ for $j \geq i$. On the basis of Theorem 4.3, Δ_S is the interior of $\overline{\Delta}_S$ for $m < n$ and the union of it and the interior of the face opposite to the vertex whose coordinates are all equal to each other for $m = n$.

We can carry all the previous results back to D_Ξ and summarize them in the following theorem.

Theorem 4.4 *If D_Ξ is the set of generic density operators with the same set Ξ of non-null eigenstates, each $\rho_S \in D_\Xi$ can be uniquely expanded as*

$$\rho_S = \sum_{\sigma \in \Xi} p_\sigma P_\sigma,$$

where P_σ is the projector associated to the eigenstate σ and the expansion coefficients p_σ are arbitrarily real different positive numbers such that $\sum_{\sigma \in \Xi} p_\sigma = 1$. With respect to the standard topology induced by the coordinates p_σ, D_Ξ consists of $m!$ connected components in bijection with the $m!$ total order relations that can be defined on Ξ. The closure $\bar{\Delta}_<$ of the connected component $\Delta_<$ corresponding to a specific order $<$ is a simplex whose vertices can be labeled with the states of Ξ. The vertex $\varepsilon^\sigma_<$ of $\bar{\Delta}_<$ is expressed as

$$\varepsilon^\sigma_< = (tr \sum_{\sigma' \geq \sigma} P_{\sigma'})^{-1} \sum_{\sigma' \geq \sigma} P_{\sigma'}.$$

The component $\Delta_<$ is the interior of $\bar{\Delta}_<$.

Remark 4.2 The positions of the vertices of $\bar{\Delta}_<$ have a simple geometric interpretation: the vertex $\varepsilon_<^\sigma$ is the barycenter of all the vertices of T_Ξ but those preceding σ. For example, if T_Ξ is a triangle whose set of vertices is $\{A, B, C\}$ and the total order $<$ is $A < B < C$, $\varepsilon_<^A$ is the barycenter of the triangle, $\varepsilon_<^B$ is the barycenter of the side BC, and $\varepsilon_<^C$ is C.

4.2 Universality of the Probability Distribution Function of a Generic Density Operator

Once the structure of D_Ξ has been thoroughly investigated, we study the problem of the probability of the non-null eigenstates of a generic density operator.

We suppose first that the probability distribution depends uniquely on the state ζ of the composite system. This is a reasonable assumption: its meaning is that the description of $\mathscr{C}$ through its state is complete, so the behavior of any of its parts is completely specified by this state. We claim that, really, it depends only on ρ_S. Indeed, if we have an evolution T represented by a unitary transformation of the form $I \otimes V$, $\mathscr{H}$ can be considered as isolated and there is no evolution in it. Consequently, the probability of any state ξ of $\mathscr{H}$ as an isolated subsystem of $\mathscr{C}$ does not change. This means that it depends only on the orbit of ζ and hence only on ρ_S. Consequently, *given the composite system and its subsystem $\mathscr{H}$, a mapping f arises which associates to each generic density operator ρ_S, the probability distribution of its non-null eigenstates.*

We will now prove the following theorem.

Theorem 4.5 *If Ξ, Ξ' are the sets of non-null eigenstates of ρ_S, ρ'_S, and* $\mathbf{p}$, $\mathbf{p}'$ *denotes the mappings which associate to each non-null eigenstate the corresponding eigenvalue then, if ρ_S and ρ'_S have the same spectrum, there is a unique bijection $\Xi \xrightarrow{\beta} \Xi'$ such that*

$$\mathbf{p}' = \mathbf{p} \circ \beta^{-1}$$

and, if $\mathbf{q}$ *and* $\mathbf{q}'$ *are the probability distributions of Ξ and Ξ', then*

$$\mathbf{q}' = \mathbf{q} \circ \beta^{-1}.$$

Proof If ρ_S and ρ'_S have the same spectrum, then $\mathrm{Im}\,\mathbf{p} = \mathrm{Im}\,\mathbf{p}'$. Furthermore, the restrictions $\widehat{\mathbf{p}}$ and $\widehat{\mathbf{p}'}$ of $\mathbf{p}$ and $\mathbf{p}'$ to their images are bijections. Hence, $(\widehat{\mathbf{p}'})^{-1} \circ \widehat{\mathbf{p}}$ is the unique required bijection β. This bijection can be extended (although in a non-unique way) to a transformation $T \in \mathscr{U}^Z(\mathscr{H})$. This means that there is an evolution of the isolated system $\mathscr{H}$ such that, whenever the state "before" is ξ and the state "after" is $\beta(\xi)$. This entails that $\mathbf{q}'(\xi') = \mathbf{q}(\beta^{-1}(\xi'))$

We now exploit Theorem 4.5 and draw from it some useful consequences.

Let Ξ be a set of m orthogonal states and D_Ξ the set of generic density operators admitting Ξ as the set of its non-null eigenstates. There is a bijection

$$D_\Xi \xrightarrow{\delta_\Xi} \Delta_\Xi$$

between D_Ξ and the set Δ_Ξ of all the injective mappings $\mathbf{p} : \Xi \xrightarrow{\mathbf{p}} \mathbb{R}_>$ such that $\sum_{\sigma \in \Xi} \mathbf{p}(\sigma) = 1$. This bijection is defined as

$$\delta_\Xi (\rho_S)(\sigma) = tr(P_\sigma \rho_S). \tag{4.2.1}$$

We have

$$\delta_\Xi^{-1}(\mathbf{p}) = \sum_{\sigma \in \Xi} \mathbf{p}(\sigma) P_\sigma. \tag{4.2.2}$$

Let f be the mapping which associates to each generic ρ_S the probability distribution of its eigenstates. We denote f_m its restriction to the set of generic density operators of type m. If f_Ξ is the restriction of f to D_Ξ, we define the mapping

$$\widehat{f_\Xi} = f_\Xi \circ \delta_\Xi^{-1}. \tag{4.2.3}$$

Let $\Delta_{\mathbb{N}_m}$ be the set of injective mappings from $\mathbb{N}_m$ to $\mathbb{R}_>$ and $\mathbb{N}_m \xrightarrow{\gamma} \Xi$ a bijection; we define a mapping through the position

$$g_\Xi^{(\gamma)}(\mathbf{m}) = \widehat{f_\Xi}\left(\mathbf{m} \circ \gamma^{-1}\right) \circ \gamma, \tag{4.2.4}$$

for $\mathbf{m} \in \Delta_{\mathbb{N}_m}$. We can prove the following.

Theorem 4.6 *The mapping $g_\Xi^{(\gamma)}$ is independent of γ.*

Proof Let β be any permutation of Ξ. For each $\mathbf{p} \in \Delta_\Xi$, we define $\rho_S = \delta_\Xi^{-1}(\mathbf{p})$ and $\rho'_S = \delta_\Xi^{-1}\left(\mathbf{p} \circ \beta^{-1}\right)$. Their spectra are $\operatorname{Im} \mathbf{p}$ and $\operatorname{Im}\left(\mathbf{p} \circ \beta^{-1}\right)$, respectively, and hence they are the same. If $\mathbf{q}$ and $\mathbf{q}'$ are the corresponding probability distributions, $\mathbf{q} = f_\Xi(\rho_S)$ and $\mathbf{q}' = f_\Xi\left(\rho'_S\right)$. Owing to Theorem 4.5 we get $f_\Xi\left(\delta_\Xi^{-1}\left(\mathbf{p} \circ \beta^{-1}\right)\right) = f_\Xi\left(\delta_\Xi^{-1}(\mathbf{p})\right) \circ \beta^{-1}$, that is,

$$\widehat{f_\Xi}\left(\mathbf{p} \circ \beta^{-1}\right) = \widehat{f_\Xi}(\mathbf{p}) \circ \beta^{-1}.$$

Now, if $\mathbb{N}_m \xrightarrow{\bar{\gamma}} \Xi$ is another bijection, $\beta \triangleq \bar{\gamma} \circ \gamma^{-1}$ is a permutation of Ξ; further, we have

$$g_\Xi^{(\bar{\gamma})}(\mathbf{m}) = \widehat{f_\Xi}\left(\mathbf{m} \circ \bar{\gamma}^{-1}\right) \circ \bar{\gamma} = \widehat{f_\Xi}\left(\mathbf{m} \circ \gamma^{-1} \circ \beta^{-1}\right) \circ \beta \circ \gamma = \widehat{f_\Xi}\left(\mathbf{m} \circ \gamma^{-1}\right) \circ \beta^{-1} \circ \beta \circ \gamma = g_\Xi^{(\gamma)}(\mathbf{m}).$$

□

We now show that

Theorem 4.7 *The mapping g_Ξ is independent of Ξ.*

Proof Let $\Xi \xrightarrow{\beta} \Xi'$ be a bijection. Applying Theorem 4.5, we get

$$\widehat{f_{\Xi'}}(\mathbf{p}) = \widehat{f_\Xi}(\mathbf{p} \circ \beta) \circ \beta^{-1}. \tag{4.2.5}$$

If $\mathbb{N}_m \xrightarrow{\gamma} \Xi'$ is a bijection, we have

$$g_{\Xi'}(\mathbf{m}) = \widehat{f_{\Xi'}}\left(\mathbf{m} \circ \gamma^{-1}\right) \circ \gamma = \widehat{f_\Xi}\left(\mathbf{m} \circ \gamma^{-1} \circ \beta\right) \circ \beta^{-1} \circ \gamma. \tag{4.2.6}$$

The mapping $\bar{\gamma} = \beta^{-1} \circ \gamma$ is a bijection $\mathbb{N}_m \xrightarrow{\bar{\gamma}} \Xi$ and

$$g_{\Xi'}(\mathbf{m}) = \widehat{f_\Xi}\left(\mathbf{m} \circ \bar{\gamma}^{-1}\right) \circ \bar{\gamma} = g_\Xi(\mathbf{m}). \tag{4.2.7}$$

□

Remark 4.3 The domain of g_Ξ is $\Delta_{\mathbb{N}_m}$. For each $\mathbf{m}$, the value of g_Ξ is a mapping which associates to each element of $\mathbb{N}_m$ the probability of the corresponding state in the set Ξ of possible states, so that this mapping is an element of the standard $(m-1)$-simplex T_m. Therefore, g_Ξ is a mapping from $\Delta_{\mathbb{N}_m}$ in T_m. We have just shown that it depends on Ξ only through its cardinality m, so that it will be denoted as g_m. The domain will be more simply denoted Δ_m, so that g_m is a mapping from Δ_m (a subset of T_m) in T_m.

Although the argumentations developed so far do not exclude a possible dependence of g_m on the system, we will next show that this is not the case. Before we pursue this goal, it is useful to furnish an explicit expression of the restriction of f to the density operators of rank m in terms of g_m. This is given by the following theorem.

Theorem 4.8 *If ρ_S is a generic density operator of rank m,*

$$f(\rho_S) = g_m(\delta_\Xi(\rho_S) \circ \gamma) \circ \gamma^{-1}$$

where Ξ is the set of eigenstates appearing in the spectral decomposition of ρ_S and $\mathbb{N}_m \xrightarrow{\gamma} \Xi$ an arbitrary bijection.

Proof With the notations previously introduced, we get

$$f(\rho_S)(\sigma) = \mathbf{q}(\sigma) = \widehat{f_\Xi}(\mathbf{p})(\sigma). \tag{4.2.8}$$

If $\mathbb{N}_m \xrightarrow{\gamma} \Xi$ is a bijection, putting $\mathbf{p} = \mathbf{m} \circ \gamma^{-1}$, we get

$$f(\rho_S)(\sigma) = \widehat{f}_{\Xi}\left(\mathbf{m} \circ \gamma^{-1}\right)\left(\gamma \circ \gamma^{-1}(\sigma)\right) = g_m(\mathbf{m}) \circ \gamma^{-1}(\sigma). \qquad (4.2.9)$$

But $\mathbf{m} = \mathbf{p} \circ \gamma$, and $\mathbf{p} = \delta_{\Xi}(\rho_S)$, so that $f_m(\rho_S)(\sigma) = g_m(\delta_{\Xi}(\rho_S) \circ \gamma) \circ \gamma^{-1}(\sigma)$. □

In the sequel, the following theorem will prove useful.

Theorem 4.9 *If π is a permutation of $\mathbb{N}_m$, then $g_m(\mathbf{m} \circ \pi) = g_m(\mathbf{m}) \circ \pi$.*

Proof In Theorem 4.8 the bijection γ is arbitrary. Hence, if $\bar{\gamma} = \gamma \circ \pi$,

$$g_m(\mathbf{m} \circ \pi) \circ \pi^{-1} \circ \gamma^{-1} = g_m(\mathbf{m}) \circ \gamma^{-1},$$

that is, $g_m(\mathbf{m} \circ \pi) = g_m(\mathbf{m}) \circ \pi$. □

Remark 4.4 The mapping g_m is defined on a subset of the standard $(m-1)$-simplex. The set of barycentric coordinates is not independent. Hence, when we write $g_m(\mathbf{m})$ we must regard $\mathbf{m}$ as a point of T_m rather than a collection of free coordinates.

In what follows, we will use for T_m a standard system of free coordinates defined as $x_i(\mathbf{m}) = \mathbf{m}(i), i \in \mathbb{N}_{m-1}$. This position yields a bijection between T_m and the subset $\widehat{T}_{m-1}$ of $\mathbb{R}^{m-1}$ defined by the inequalities $x_i \geq 0$, $\sum_{i=1}^{m-1} x_i \leq 1$. If $T_m \xrightarrow{\eta} \widehat{T}_{m-1}$ is this bijection, in standard coordinates, we use the mapping $\widehat{g}_m = \eta \circ g_m \eta^{-1}$. The domain Δ_m of g_m is given by all the sets of different positive barycentric coordinates. Hence, the domain $\widehat{\Delta}_m$ of $\widehat{g}_m$ is given by all the sets of different standard coordinates in $\widehat{T}_{m-1}$, with the further conditions $1 - \sum_{j=1}^{m-1} x_j \neq x_i, i \in \mathbb{N}_{m-1}$, and excluding further the boundary of $\widehat{T}_{m-1}$.

We now pursue the goal of showing that g_m *does not depend on the system* (provided that the dimension of its Hilbert space is at least m).

To this purpose, let us envision a composite system $\mathscr{S}\mathscr{S}'$ described by a Hilbert space $\mathscr{H} \otimes \mathscr{K}$. Suppose that this system interacts with an environment $\mathscr{E}$ described by a Hilbert space $\mathscr{L}$, so that the total system is described by $\mathscr{H} \otimes \mathscr{K} \otimes \mathscr{L}$. Suppose that initially the system is in a state described by a ket $|\phi\rangle\,|\kappa\rangle\,|\theta\rangle$.

Assume that then an interaction between $\mathscr{S}'$ and $\mathscr{E}$ takes place, giving rise to a state $|\phi\rangle\left(\sum d_\alpha\,|\kappa_\alpha\rangle\,|\theta_\alpha\rangle\right)$ (where the term in parentheses is the canonical decomposition of the final state of $\mathscr{S}'\mathscr{E}$) with the coefficients d_α all different to each other and in number of l.

The number l and the quantities d_α can be chosen at will provided that $l \leq \dim \mathscr{K}$, and that the dimension of $\mathscr{L}$ is chosen not smaller than that of $\mathscr{K}$ and that $\sum d_\alpha^2 = 1$.

Finally, suppose to have an interaction between $\mathscr{S}'$ and $\mathscr{S}$ that brings from each $|\phi\rangle\,|\kappa_\alpha\rangle$ to $|\phi_\alpha\rangle\,|\kappa\rangle$, where the $|\phi_\alpha\rangle$ form an orthonormal set of $\mathscr{H}$. This is certainly possible if $l \leq \dim \mathscr{H}$ because the $|\phi\rangle\,|\kappa_\alpha\rangle$, as well as the $|\phi_\alpha\rangle\,|\kappa\rangle$, are orthonormal sets in $\mathscr{H} \otimes \mathscr{K}$. It brings the total system from the state $\sum d_\alpha |\phi\rangle |\kappa_\alpha\rangle |\theta_\alpha\rangle$ to the state $\sum d_\alpha |\phi_\alpha\rangle |\kappa\rangle |\theta_\alpha\rangle$.

Consider $\mathscr{S}\mathscr{S}'$ as a subsystem of $\mathscr{S}\mathscr{S}'\mathscr{E}$. The initial density operator is

$$\rho_{\mathscr{S}\mathscr{S}'} = \sum d_\alpha^2 |\phi\rangle \langle\phi| \otimes |\kappa_\alpha\rangle \langle\kappa_\alpha|. \tag{4.2.10}$$

The set of non-null eigenstates of $\rho_{\mathscr{S}\mathscr{S}'}$ is $X = \{|\phi\rangle |\kappa_\alpha\rangle\}$. Whenever $\mathscr{S}\mathscr{S}'$ is in the state $|\phi\rangle |\kappa_\alpha\rangle$, $\mathscr{S}$ is in $|\phi\rangle$ and $\mathscr{S}'$ is in $|\kappa_\alpha\rangle$ and hence $\mathscr{S}'$ is in $|\kappa_\alpha\rangle$. *Vice versa*, whenever $\mathscr{S}'$ is in $|\kappa_\alpha\rangle$, $\mathscr{S}\mathscr{S}'$ is in $|\phi\rangle |\kappa_\alpha\rangle$ so that $\Pr_{\mathscr{S}\mathscr{S}'} |\phi\rangle |\kappa_\alpha\rangle = \Pr_{\mathscr{S}'} |\kappa_\alpha\rangle$. Let us use the elements of $\mathbb{N}_l$ to label the states of X, so that we define

$$\mathbb{N}_l \xrightarrow{\gamma} X : \gamma(\alpha) = |\phi\rangle |\kappa_\alpha\rangle. \tag{4.2.11}$$

We get $\gamma^{-1}(|\phi\rangle |\kappa_\alpha\rangle) = \alpha$. Furthermore, $\delta_X(\rho_{\mathscr{S}\mathscr{S}'}) \circ \gamma(\alpha) = d_\alpha^2$. Hence, if $\mathbf{m}(\alpha) \triangleq d_\alpha^2$, we get $\Pr_{\mathscr{S}\mathscr{S}'} |\phi\rangle |\kappa_\alpha\rangle = g_l{}^{\mathscr{S}\mathscr{S}'}(\mathbf{m})(\alpha)$.

In order to express $\Pr_{\mathscr{S}'} |\kappa_\alpha\rangle$, we must evaluate $\rho_{\mathscr{S}'}$. We have $\rho_{\mathscr{S}\mathscr{S}'} = \sum d_\alpha^2 |\kappa_\alpha\rangle \langle\kappa_\alpha|$. With the same procedure we find $\Pr_{\mathscr{S}'} |\kappa_\alpha\rangle = g_l{}^{\mathscr{S}'}(\mathbf{m})(\alpha)$. Hence, in this case, $g_l{}^{\mathscr{S}'}(\mathbf{m})(\alpha) = g_l{}^{\mathscr{S}\mathscr{S}'}(\mathbf{m})(\alpha)$. Similarly, the final density operator is

$$\rho'_{\mathscr{S}\mathscr{S}'} = \sum d^2{}_\alpha |\phi_\alpha\rangle \langle\phi_\alpha| \otimes |\kappa\rangle \langle\kappa|. \tag{4.2.12}$$

Using once again the same method, we conclude that $g_l{}^{\mathscr{S}}(\mathbf{m})(\alpha) = g_l{}^{\mathscr{S}\mathscr{S}'}(\mathbf{m})(\alpha)$. Hence $g_l{}^{\mathscr{S}}(\mathbf{m}) = g_l{}^{\mathscr{S}'}(\mathbf{m})$. Owing to the previous considerations, $\mathbf{m}$ is an arbitrary point of Δ_l, so that $g_{l\mathscr{S}} = g_{l\mathscr{S}'}$.

We can express the results just obtained in the following.

Theorem 4.10 *For each m, a universal mapping g_m exists such that in every system supporting density operators of rank m the probability law on the set Ξ of their possible states is expressed as $f(\rho_S) = g_m(\delta_\Xi(\rho_S) \circ \gamma) \circ \gamma^{-1}$.*

4.3 A Functional Equation for g_2

Our next task is the determination of g. We will first determine g_2, and then we will find it for a general rank m. As for the moment, our analysis is focused on $m = 2$, we will write it simply g. In this section, we will find a functional equation which must be obeyed by g. The next section will be dedicated to its solution.

Let us consider a system $\mathscr{W}$ composed of three systems $\mathscr{R}$, $\mathscr{S}$, and $\mathscr{T}$. We call $\mathscr{H}$, $\mathscr{K}$, and $\mathscr{L}$ the Hilbert spaces of $\mathscr{R}$, $\mathscr{S}$, and $\mathscr{T}$, respectively, and $\mathscr{M} = \mathscr{H} \otimes \mathscr{K} \otimes \mathscr{L}$ the Hilbert space of $\mathscr{W}$. We suppose that $\mathscr{H}$, $\mathscr{K}$, and $\mathscr{L}$ are two-dimensional. It is useful for what follows to introduce some special states of $\mathscr{W}$ that will be called distinguished states. In order to simplify the terminology, we will say "state $|\chi\rangle$" instead of "state specified by the ket $|\chi\rangle$." Furthermore, all the kets will be normalized.

Definition 4.1 We say that a ket $|\chi\rangle \in \mathscr{M}$ is a distinguished state if the following conditions are satisfied.

(a) The partial density operator ρ_L is generic and with nonzero eigenvalues;
(b) In the canonical representation, $|\chi\rangle = d_1 |\psi_1\rangle |l_1\rangle + d_2 |\psi_2\rangle |l_2\rangle$ with $d_1 \neq d_2$,$d_1 > 0$, $d_2 > 0$, $|l_1\rangle, |l_2\rangle$ orthonormal vectors of $\mathscr{L}$ and $|\psi_1\rangle, |\psi_2\rangle$ orthonormal vectors of $\mathscr{H} \otimes \mathscr{K}$, $tr_K |\psi_1\rangle \langle\psi_1|$ and $tr_K |\psi_2\rangle \langle\psi_2|$ are generic and have the same eigenstates.

We can show that

Theorem 4.11 *If* $|\chi\rangle = d_1 |\psi_1\rangle |l_1\rangle + d_2 |\psi_2\rangle |l_2\rangle$ *with* $d_1 \neq d_2$, $d_1 > 0$, $d_2 > 0$, $|l_1\rangle, |l_2\rangle$ *orthonormal vectors of* $\mathscr{L}$ *and*

$$|\psi_1\rangle = \sqrt{m_1}\, |h_1\rangle |k_1\rangle + \sqrt{m_2}\, |h_2\rangle |k_2\rangle, \tag{4.3.1}$$

$$|\psi_2\rangle = \sqrt{m'_1}\, |h_1\rangle |k_2\rangle + \sqrt{m'_2}\, |h_2\rangle |k_1\rangle, \tag{4.3.2}$$

with $m_i, m'_i \geq 0$, $m_1 + m_2 = 1$, $m'_1 + m'_2 = 1$, $m_1 \neq m_2$, $m'_1 \neq m'_2$, $|h_1\rangle$, $|h_2\rangle$ *orthonormal in* $\mathscr{H}$, $|k_1\rangle, |k_2\rangle$ *orthonormal in* $\mathscr{K}$, *then* $|\chi\rangle$ *is a distinguished state.*

Proof $|\psi_1\rangle$ and $|\psi_2\rangle$ are orthonormal; ρ_L is generic and with nonzero eigenvalues as a consequence of the assumptions on d_1 and d_2. Furthermore

$$tr_K |\psi_1\rangle \langle\psi_1| = m_1 |h_1\rangle \langle h_1| + m_2 |h_2\rangle \langle h_2|, \tag{4.3.3}$$

and

$$tr_K |\psi_2\rangle \langle\psi_2| = m'_1 |h_1\rangle \langle h_1| + m'_2 |h_2\rangle \langle h_2|, \tag{4.3.4}$$

so that they are generic and with the same eigenstates. □

We have $\rho_{HK} = d_1^2 |\psi_1\rangle \langle\psi_1| + d_2^2 |\psi_2\rangle \langle\psi_2|$. Putting $\lambda = d_1^2$, $\lambda' = d_2^2$, $\rho_1 = tr_K |\psi_1\rangle \langle\psi_1|$ and $\rho_2 = tr_K |\psi_2\rangle \langle\psi_2|$, we get

$$\rho_{HK} = \lambda |\psi_1\rangle \langle\psi_1| + \lambda' |\psi_2\rangle \langle\psi_2| \text{ and } \rho_H = \lambda\rho_1 + \lambda'\rho_2.$$

The possible states of the system $\mathscr{RS}$ are $|\psi_1\rangle$ and $|\psi_2\rangle$. We get

$$\rho_H = \left(\lambda m_1 + \lambda' m'_1\right) |h_1\rangle \langle h_1| + \left(\lambda m_2 + \lambda' m'_2\right) |h_2\rangle \langle h_2|,$$

so that, if $\lambda m_1 + \lambda' m'_1 \neq \frac{1}{2}$, the system $\mathscr{R}$ can be considered isolated in ρ_H. Hence, as long as neither $m_1 + m'_1$ nor $m_2 + m'_2$ are zero, when the total system is in $|\chi\rangle$, $\mathscr{R}$ is either in $|h_1\rangle$ or in $|h_2\rangle$.

Suppose first that all the m and m' are nonzero. In this case, each possibility for the states of $\mathscr{RS}$ as an isolated system, i.e., $|\psi_1\rangle$ and $|\psi_2\rangle$, gives rise to both possibilities for the states of $\mathscr{R}$, that is, $|h_1\rangle$ and $|h_2\rangle$. Hence, we have two disjoint possibilities of obtaining $|h_1\rangle$: either with $\mathscr{RS}$ in $|\psi_1\rangle$ or with $\mathscr{RS}$ in $|\psi_2\rangle$. Consequently, the probability $\bar{q}_1$ of finding $\mathscr{R}$ in $|h_1\rangle$ is the sum of two terms corresponding to these possibilities. Applying the Bayes rule, the first term is the product of the probability q_1 of finding $\mathscr{RS}$ in $|\psi_1\rangle$ times the probability $q_1^{(1)}$ of finding $\mathscr{R}$ in $|h_1\rangle$ given that $\mathscr{RS}$ is in $|\psi_1\rangle$. Similarly, the second term is the product of the probability q_2 of finding $\mathscr{RS}$ in $|\psi_2\rangle$ times the probability $q_1^{(2)}$ of finding $\mathscr{R}$ in $|h_1\rangle$ given that $\mathscr{RS}$ is in $|\psi_2\rangle$. Hence, we can write

$$\bar{q}_1 = q_1^{(1)} q_1 + q_1^{(2)} q_2. \tag{4.3.5}$$

In the same way, we get

$$\bar{q}_2 = q_2^{(1)} q_1 + q_2^{(2)} q_2, \tag{4.3.6}$$

where $q_2^{(1)}$ is the probability of finding $\mathscr{R}$ in $|h_2\rangle$ given that $\mathscr{RS}$ is in $|\psi_1\rangle$, and $q_2^{(2)}$ is the probability of finding $\mathscr{R}$ in $|h_2\rangle$ given that $\mathscr{RS}$ is in $|\psi_2\rangle$.

Denoting with $g_{(1)}$ and $g_{(2)}$ the two components of g, we have

$$\begin{aligned} q_1 &= g_{(1)}\left(\lambda, \lambda'\right), & q_2 &= g_{(2)}\left(\lambda, \lambda'\right), \\ q_1^{(1)} &= g_{(1)}\left(m_1, m_2\right), & q_2^{(1)} &= g_{(2)}\left(m_1, m_2\right), \\ q_1^{(2)} &= g_{(1)}\left(m'_1, m'_2\right), & q_2^{(2)} &= g_{(2)}\left(m'_1, m'_2\right). \end{aligned}$$

Furthermore

$$\bar{q}_1 = g_{(1)}\left(\lambda m_1 + \lambda' m'_1, \lambda m_2 + \lambda' m'_2\right), \tag{4.3.7}$$
$$\bar{q}_2 = g_{(2)}\left(\lambda m_1 + \lambda' m'_1, \lambda m_2 + \lambda' m'_2\right). \tag{4.3.8}$$

By substitution, we get

$$\begin{aligned} g_{(1)}\left(\lambda m_1 + \lambda' m'_1, \lambda m_2 + \lambda' m'_2\right) &= g_{(1)}\left(m_1, m_2\right) g_{(1)}\left(\lambda, \lambda'\right) + g_{(1)}\left(m'_1, m'_2\right) g_{(2)}\left(\lambda, \lambda'\right), \\ g_{(2)}\left(\lambda m_1 + \lambda' m'_1, \lambda m_2 + \lambda' m'_2\right) &= g_{(2)}\left(m_1, m_2\right) g_{(1)}\left(\lambda, \lambda'\right) + g_{(2)}\left(m'_1, m'_2\right) g_{(2)}\left(\lambda, \lambda'\right). \end{aligned}$$

These equations hold for every $\left(\left(\lambda, \lambda'\right), \mathbf{m}, \mathbf{m}'\right) \in \Delta_2^3$ (where $\mathbf{m} = (m_1, m_2)$, $\mathbf{m}' = (m'_1, m'_2)$) such that $\lambda\mathbf{m} + \lambda'\mathbf{m}' \in \Delta_2$. We now take $m_1 = 1$, maintaining the condition that neither $m_1 + m'_1$ nor $m_2 + m'_2$ are zero. In this case, the possible state $|\psi_1\rangle$ gives rise to the single possibility $|h_1\rangle$. Further m'_2 must be nonzero. If in addition $m'_2 \neq 1$, $|\psi_2\rangle$ gives rise to the both possibilities $|h_1\rangle$ and $|h_2\rangle$. Hence, in this case, we must write a single summand for $\bar{q}_2$ and identify with q_1 the summand corresponding to the possibility $|\psi_1\rangle$ in $\bar{q}_1$; this is equivalent to take $q_1^{(1)} = 1$ and $q_2^{(1)} = 0$. So, we find

$$\bar{q}_1 = q_1 + q_1^{(2)} q_2, \tag{4.3.9}$$

$$\bar{q}_2 = q_2^{(2)} q_2, \tag{4.3.10}$$

giving rise to the equations

$$g_{(1)}\left(\lambda + \lambda' m'_1, \lambda' m'_2\right) = g_{(1)}\left(\lambda, \lambda'\right) + g_{(1)}\left(m'_1, m'_2\right) g_{(2)}\left(\lambda, \lambda'\right), \tag{4.3.11}$$

$$g_{(2)}\left(\lambda + \lambda' m'_1, \lambda' m'_2\right) = g_{(2)}\left(m'_1, m'_2\right) g_{(2)}\left(\lambda, \lambda'\right). \tag{4.3.12}$$

If $m_1 = 1, m'_2 = 1$, $|\psi_2\rangle$ gives rise to the single possibility $|h_2\rangle$. In this case, an identity is obtained. We now take $m_2 = 1$ and m'_1, m'_2 both nonzero. This time we get $q_1^{(1)} = 0$ and $q_2^{(1)} = 1$ and hence

$$\bar{q}_1 = q_1^{(2)} q_2, \tag{4.3.13}$$

$$\bar{q}_2 = q_1 + q_2^{(2)} q_2, \tag{4.3.14}$$

so that

$$g_{(1)}\left(\lambda' m'_1, \lambda + \lambda' m'_2\right) = g_{(1)}\left(m'_1, m'_2\right) g_{(2)}\left(\lambda, \lambda'\right), \tag{4.3.15}$$

$$g_{(2)}\left(\lambda' m'_1, \lambda + \lambda' m'_2\right) = g_{(1)}\left(\lambda, \lambda'\right) + g_{(2)}\left(m'_1, m'_2\right) g_{(2)}\left(\lambda, \lambda'\right). \tag{4.3.16}$$

Similarly, we take $m'_1 = 1$ and then $m'_2 = 1$. In the first case, we get $q_1^{(2)} = 1$ and $q_2^{(2)} = 0$, in the second $q_1^{(2)} = 0$ and $q_2^{(2)} = 1$. We obtain, respectively,

$$g_{(1)}\left(\lambda m_1 + \lambda', \lambda m_2\right) = g_{(1)}\left(m_1, m_2\right) g_{(1)}\left(\lambda, \lambda'\right) + g_{(2)}\left(\lambda, \lambda'\right), \tag{4.3.17}$$

$$g_{(2)}\left(\lambda m_1 + \lambda', \lambda m_2\right) = g_{(2)}\left(m_1, m_2\right) g_{(1)}\left(\lambda, \lambda'\right), \tag{4.3.18}$$

and

$$g_{(1)}\left(\lambda m_1, \lambda m_2 + \lambda'\right) = g_{(1)}\left(m_1, m_2\right) g_{(1)}\left(\lambda, \lambda'\right), \tag{4.3.19}$$

$$g_{(2)}\left(\lambda m_1, \lambda m_2 + \lambda'\right) = g_{(2)}\left(m_1, m_2\right) g_{(1)}\left(\lambda, \lambda'\right) + g_{(2)}\left(\lambda, \lambda'\right). \tag{4.3.20}$$

We now *define* an extension $\tilde{g}$ of g as follows:

$$\tilde{g}\left(m_1, m_2\right) = g\left(m_1, m_2\right) \quad \text{for} \quad \left(m_1, m_2\right) \in \Delta_2;$$

$$\tilde{g}_{(1)}(1, 0) = 1, \tilde{g}_{(2)}(1, 0) = 0, \tilde{g}_{(1)}(0, 1) = 0, \tilde{g}_{(2)}(0, 1) = 1.$$

Defining

$$\tilde{\Delta}_2 = \Delta_2 \cup \{(1, 0), (0, 1)\},$$

we can show that $\tilde{g}$ satisfies the equations

$$\tilde{g}_{(1)}\left(\lambda m_1+\lambda' m'_1, \lambda m_2+\lambda' m'_2\right)=\tilde{g}_{(1)}\left(m_1, m_2\right) \tilde{g}_{(1)}\left(\lambda, \lambda'\right)+\tilde{g}_{(1)}\left(m'_1, m'_2\right) \tilde{g}_{(2)}\left(\lambda, \lambda'\right)$$
$$\tilde{g}_{(2)}\left(\lambda m_1+\lambda' m'_1, \lambda m_2+\lambda' m'_2\right)=\tilde{g}_{(2)}\left(m_1, m_2\right) \tilde{g}_{(1)}\left(\lambda, \lambda'\right)+\tilde{g}_{(2)}\left(m'_1, m'_2\right) \tilde{g}_{(2)}\left(\lambda, \lambda'\right)$$

for every $\left(\left(\lambda, \lambda'\right), \mathbf{m}, \mathbf{m}'\right) \in \tilde{\Delta}_2^3$ such that $\lambda \mathbf{m}+\lambda' \mathbf{m}' \in \tilde{\Delta}_2$.

Indeed, it is sufficient to verify the above equations when at least one point is $(1,0)$ or $(0,1)$. If $\left(\lambda, \lambda'\right)=(1,0)$ or $(0,1)$, we obtain an identity. The other cases are a consequence of equations already established for the remaining four possibilities.

Let us express this relation in free coordinates. We put $\mathbf{m}=\eta^{-1}(z)$ and $\mathbf{m}'=\eta^{-1}\left(z'\right)$ so that

$$\lambda \mathbf{m}+\lambda' \mathbf{m}'=\eta^{-1}\left(\lambda z+(1-\lambda) z'\right). \tag{4.3.21}$$

Therefore

$$\eta \circ \tilde{g}\left(\lambda \mathbf{m}+\lambda' \mathbf{m}'\right)=\tilde{g}_{(1)}\left(\lambda z+(1-\lambda) z'\right) \triangleq \hat{g}\left(\lambda z+(1-\lambda) z'\right).$$

We further have

$$\eta \circ \tilde{g}(\mathbf{m})=\hat{g}(z) \quad \text { and } \quad \eta \circ \tilde{g}\left(\mathbf{m}'\right)=\hat{g}\left(z'\right);$$

finally,

$$\tilde{g}_{(1)}\left(\lambda, \lambda'\right)=\hat{g}(\lambda) \quad \text { and } \quad \tilde{g}_{(2)}\left(\lambda, \lambda'\right)=1-\hat{g}(\lambda).$$

Hence, we get

$$\hat{g}\left(\lambda z+(1-\lambda) z'\right)=\hat{g}(\lambda) \hat{g}(z)+(1-\hat{g}(\lambda)) \hat{g}\left(z'\right), \tag{4.3.22}$$

which holds for all the $\left(\lambda, z, z'\right)$ such that the arguments in all the occurrences of $\hat{g}$ belong to its domain $[0,1/2) \bigcup (1/2,1]$.

The definition of $\tilde{g}$ entails that the boundary conditions $\hat{g}(0)=0$ and $\hat{g}(1)=1$ must be satisfied.

The next section will be dedicated to the solution of this functional equation.

4.4 Solution of the Functional Equation

In the previous section, we have found the functional equation

$$\hat{g}\left(\lambda z+(1-\lambda) z'\right)=\hat{g}(\lambda) \hat{g}(z)+(1-\hat{g}(\lambda)) \hat{g}\left(z'\right). \tag{4.4.1}$$

The function $\hat{g}$ is defined in $\Delta_2=\left[0, \frac{1}{2}\right) \bigcup\left(\frac{1}{2}, 1\right]$. Furthermore, the boundary conditions $\hat{g}(0)=0$, $\hat{g}(1)=1$ must be fulfilled and its values must range between 0 and 1. This equation holds in Δ_2^3 for $\lambda z+(1-\lambda) z' \neq \frac{1}{2}$.

This section is dedicated to its solution. First, we make the following positions:

$$x = z - \frac{1}{2}, x' = z' - \frac{1}{2}, \mu = \lambda - \frac{1}{2}, f(x) = \hat{g}(z) - \frac{1}{2}. \tag{4.4.2}$$

We have $\lambda z + (1-\lambda) z' = \lambda x + (1-\lambda) x' + \frac{1}{2}$, so that

$$f(\lambda x + (1-\lambda) x') + \frac{1}{2} = \left(f(\mu) + \frac{1}{2}\right)\left(f(x) + \frac{1}{2}\right) + \left(\frac{1}{2} - f(\mu)\right)\left(f(x') + \frac{1}{2}\right), \tag{4.4.3}$$

whence

$$f\left[\left(\mu + \frac{1}{2}\right)x + \left(\frac{1}{2} - \mu\right)x'\right] = f(\mu)\left(f(x) - f(x')\right) + \frac{1}{2}\left(f(x) + f(x')\right). \tag{4.4.4}$$

Putting $D = \left[-\frac{1}{2}, 0\right) \bigcup \left(0, \frac{1}{2}\right]$, this equation holds in D^3 for $\left(\mu + \frac{1}{2}\right) x + \left(\frac{1}{2} - \mu\right) x' \neq 0$. In particular, for $x' = -x$, we get

$$f(2\mu x) = f(\mu)(f(x) - f(-x)) + \frac{1}{2}(f(x) + f(-x)). \tag{4.4.5}$$

This equation holds in D^2.

The left side is symmetric under the exchange of μ and x, so that we have

$$-f(\mu) f(-x) + \frac{1}{2}(f(x) + f(-x)) = -f(x) f(-\mu) + \frac{1}{2}(f(\mu) + f(-\mu)). \tag{4.4.6}$$

We have $f\left(-\frac{1}{2}\right) = -\frac{1}{2}$ and $f\left(\frac{1}{2}\right) = \frac{1}{2}$, so that, putting $\mu = -\frac{1}{2}$, we get

$$f(x) + f(-x) = 0, \tag{4.4.7}$$

and hence

$$f(2\mu x) = 2f(\mu) f(x). \tag{4.4.8}$$

We put $F(\zeta) = 2f\left(\frac{1}{2}\zeta\right)$, so that $f(\zeta) = \frac{1}{2}F(2\zeta)$. Hence, we get $F(4\mu x) = F(2\mu) F(2x)$, that is,

$$F(xy) = F(x) F(y). \tag{4.4.9}$$

As the domain of f is D, the domain of F is $\Lambda = [-1, 0) \bigcup (0, 1]$. Equation (4.4.9) holds for any $(x, y) \in \Lambda \times \Lambda$. The functional equation for f can be translated into a functional equation for F. We have

$$\frac{1}{2}F((2\mu + 1)x + (1 - 2\mu)x') = \frac{1}{4}F(2\mu)\left(F(2x) - F(2x')\right) + \frac{1}{4}\left(F(2x) + F(2x')\right). \tag{4.4.10}$$

Putting $2\mu = \xi$, $2x = u$, $2x' = v$, we obtain

$$F\left[\frac{1}{2}(1+\xi)u+\frac{1}{2}(1-\xi)v\right]=\frac{1}{2}F(\xi)\left[F(u)-F(v)\right]+\frac{1}{2}\left[F(u)+F(v)\right], \tag{4.4.11}$$

which holds for any $(\xi, u, v) \in \Lambda^3$ such that $u+v+\xi u-\xi v \neq 0$. We can write

$$F\left(\frac{1}{2}u+\frac{1}{2}\xi u+\frac{1}{2}v-\frac{1}{2}\xi v\right)=\frac{1}{2}F(u)+\frac{1}{2}F(\xi u)+\frac{1}{2}F(v)-\frac{1}{2}F(\xi v). \tag{4.4.12}$$

We restrict the above equation to the subset Σ of Λ^3 satisfying the condition $u=\xi v$ and such that $v+\xi u \neq 0$. In this subset, we have

$$F\left(\frac{1}{2}\xi u+\frac{1}{2}v\right)=\frac{1}{2}F(\xi u)+\frac{1}{2}F(v). \tag{4.4.13}$$

We now put $\alpha=\xi u=\xi^2 v$, $\beta=v$. The inverse transformation is $\xi=\pm\sqrt{\alpha/\beta}$, $u=\pm\sqrt{\alpha\beta}$, $v=\beta$. Hence, we have $0<|\beta|\leq 1, 0<\alpha/\beta\leq 1, 0<\alpha\beta\leq 1$. This subset can be characterized by the inequalities $0<|\alpha|\leq|\beta|\leq 1$ and $\alpha\beta>0$. Therefore, the equation

$$F\left(\frac{1}{2}\alpha+\frac{1}{2}\beta\right)=\frac{1}{2}F(\alpha)+\frac{1}{2}F(\beta) \tag{4.4.14}$$

holds for $0<|\alpha|\leq 1, 0<|\beta|\leq 1$, $|\alpha|\leq|\beta|$ such that $\alpha\beta>0$. If $|\alpha|>|\beta|$ we put $\alpha'=\beta$, $\beta'=\alpha$, so that $|\alpha'|<|\beta'|$. We have

$$F\left(\frac{1}{2}\alpha+\frac{1}{2}\beta\right)=F\left(\frac{1}{2}\alpha'+\frac{1}{2}\beta'\right)=\frac{1}{2}F(\alpha')+\frac{1}{2}F(\beta')=\frac{1}{2}F(\alpha)+\frac{1}{2}F(\beta).$$

Hence Eq. (4.4.14) holds for all (α, β) such that $0<|\alpha|\leq 1, 0<|\beta|\leq 1, \alpha\beta>0$.

As $F(x)+F(-x)=0$, we can restrict our analysis to the interval $(0, 1]$. Hence, we have the equations

$$F(xy)=F(x)F(y), \tag{4.4.15}$$

$$F\left(\frac{x+y}{2}\right)=\frac{1}{2}F(x)+\frac{1}{2}F(y), \tag{4.4.16}$$

with $(x, y) \in (0, 1]^2$.

Let $X_N=\left\{\frac{n}{2^N} | n \in \mathbb{N}, n<2^N\right\}$. We have the following.

Theorem 4.12 *If F satisfies Eq. (4.4.16), for any $N \in \mathbb{N}$ and for any $x \in X_N$, $F(x)$ is given by*

$$F(x)=F\left(\frac{1}{2}\right)+\left[1-F\left(\frac{1}{2}\right)\right](2x-1). \tag{4.4.17}$$

Proof The theorem is true for $N=1$. Then we prove it by recursion on N.

If the theorem is true for a given N, all the points $(x, F(x))$ such that $x \in X_N$ lie on the straight line r containing the points $\left(\frac{1}{2}, F\left(\frac{1}{2}\right)\right)$ and $(1, 1)$.

The set X_{N+1} is obtained from X_N by adjoining to it all the middle points of the intervals $\left(\frac{n}{2^N}, \frac{n+1}{2^N}\right)$ for $1 \leq n < 2^N - 1$ and further the two points $\frac{1}{2^N} - \frac{1}{2^{N+1}}$ and $\frac{2^N-1}{2^N} + \frac{1}{2^{N+1}}$.

If $N = 1$, $X_1 = \left\{\frac{1}{2}\right\}$ and X_2 is obtained by adjunction of the points $\frac{1}{4}$ and $\frac{3}{4}$. Equation (4.4.16) shows that $F\left(\frac{3}{4}\right)$ is obtained by linear interpolation of the values $F\left(\frac{1}{2}\right)$ and $F(1) = 1$, so that $\left(\frac{3}{4}, F\left(\frac{3}{4}\right)\right)$ belongs to r. The same equation shows that the linear interpolation between the values $F\left(\frac{1}{4}\right)$ and $F\left(\frac{3}{4}\right)$ must yield $F\left(\frac{1}{2}\right)$, so that $\left(\frac{1}{4}, F\left(\frac{1}{4}\right)\right)$ belongs to r.

For $N > 1$, the value of F in the middle point of each interval $\left(\frac{n}{2^N}, \frac{n+1}{2^N}\right)$ is obtained by linear interpolation of the values in its extreme points, so that $(x, F(x))$ lies on r. For the point adjoined on the left, $F\left(\frac{1}{2^N} - \frac{1}{2^{N+1}}\right)$ must be such that the linear interpolation of it with $F\left(\frac{1}{2^N} + \frac{1}{2^{N+1}}\right)$ yields $F\left(\frac{1}{2^N}\right)$. Hence, $\left(\frac{1}{2^N} - \frac{1}{2^{N+1}}, F\left(\frac{1}{2^N} - \frac{1}{2^{N+1}}\right)\right)$ belongs to r. Similarly, $F\left(\frac{2^N-1}{2^N} + \frac{1}{2^{N+1}}\right)$ must be such that the linear interpolation of it with $F\left(\frac{2^N-1}{2^N} - \frac{1}{2^{N+1}}\right)$ yields $F\left(\frac{2^N-1}{2^N}\right)$.

Therefore $\left(\frac{2^N-1}{2^N} + \frac{1}{2^{N+1}}, F\left(\frac{2^N-1}{2^N} + \frac{1}{2^{N+1}}\right)\right)$ belongs to r. □

We now use Eq. (4.4.15) for $x = y = \frac{1}{2}$. We get $F\left(\frac{1}{4}\right) = F\left(\frac{1}{2}\right)^2$. But $\frac{1}{4} \in X_2$, so that we can use Theorem 4.12 getting $F\left(\frac{1}{4}\right) = \frac{3}{2}F\left(\frac{1}{2}\right) - \frac{1}{2}$. Hence, $F\left(\frac{1}{2}\right)$ satisfies the equation $F\left(\frac{1}{2}\right)^2 - \frac{3}{2}F\left(\frac{1}{2}\right) + \frac{1}{2} = 0$ whose solutions are $F\left(\frac{1}{2}\right) = 1$ and $F\left(\frac{1}{2}\right) = \frac{1}{2}$. In the first case, we get $F(x) = 1$ and in the second $F(x) = x$. The first case can be excluded. Indeed, as F is odd, $F(x) = sign(x)$. If we take, for instance, $\xi > 0$ $u > 0$ and such that $\frac{(1+\xi)u}{1-\xi} < 1$ and $v < -\frac{(1+\xi)u}{1-\xi}$, with the further condition that all the variables belong to $-X_N \bigcup X_N$ for some N, the argument of F on the left side is negative, so that the latter is -1. On the other hand, the right side is $F(u)$, and hence its value is 1. We conclude that the only possibility is $F(x) = x$. In this way, we have proved the following theorem.

Theorem 4.13 *For every $N \in \mathbb{N}$, the restriction of F to the set $-X_N \bigcup X_N$ is the identity.*

We are now ready to prove the next theorem.

Theorem 4.14 *The only function F satisfying Eqs. (4.4.9) and (4.4.14) and not greater than* 1 *in magnitude is the identity.*

Proof As F is an odd function, we can limit our analysis to the interval $(0, 1]$ of the independent variable.

First we observe that in this range F is nonnegative. Indeed, if $z \in (0, 1]$, and $x = \sqrt{z}$, $x \in (0, 1]$ and, by virtue of (4.4.9), we have

$$F(z) = F\left(x^2\right) = F(x)^2 \geq 0.$$

Further, F is a non-decreasing function. Indeed, if $x < y$,

$$F(x) = F\left(\left(\frac{x}{y}\right)y\right) = F\left(\frac{x}{y}\right)F(y) \leq F(y).$$

Let $z \in (0, 1]$. If $z \in X_N$ for some N, $F(z) = z$ by Theorem 4.13. If $z \notin X_N$ for every N, for each N we define x_N as the greatest $x \in X_N$ smaller than z and x'_N as the smallest $x \in X_N$ greater than z. It is obvious that $\{x_N\}$ is a non-decreasing sequence and $\{x'_N\}$ is a non-increasing sequence. Let $\bar{x} = \lim_{N\to\infty} x_N$. We have $\bar{x} \leq z$. If $\bar{x} < z$, for each N we define $\bar{x}_N$ as the smallest $x \in X_N$ greater than $\bar{x}$. The neighbor on the left of $\bar{x}_N$ is $\bar{x}_N - \frac{1}{2^N}$ which, by definition of $\bar{x}_N$, is not greater than $\bar{x}$. Hence, for sufficiently high values of N we have $\bar{x}_N < z$. But we have $x_N \leq \bar{x} < \bar{x}_N < z$. This is a contradiction because x_N is the greatest $x \in X_N$ smaller than z. Therefore, we have

$$\lim_{N\to\infty} x_N = z.$$

In a similar way, we can show that

$$\lim_{N\to\infty} x'_N = z.$$

As $x_N = F(x_N) \leq F(z) \leq F(x'_N) = x'_N$, $F(z) = z$. □

Remembering the changes of variables we have made to go from $\tilde{g}$ to F, we conclude that the unique solution of the functional equation with the conditions imposed is the identity mapping.

Remark 4.5 We emphasize that no continuity assumption has been made in the above argument. The only assumption on g besides the fact that it must satisfy the functional equation is that, as a probability, its values must range between 0 and 1.

4.5 The Function *g* for Arbitrary Rank

In this section, we generalize the results obtained before, and find the form of the function g in the general case. To this purpose, we introduce the general notion of a *distinguished state*.

Let us consider a system $\mathscr{W}$ composed of three systems $\mathscr{R}$, $\mathscr{S}$, and $\mathscr{T}$. We call $\mathscr{H}$, $\mathscr{K}$, and $\mathscr{L}$ the Hilbert spaces of $\mathscr{R}$, $\mathscr{S}$, and $\mathscr{T}$, respectively, and $\mathscr{M} = \mathscr{H} \otimes \mathscr{K} \otimes \mathscr{L}$ the Hilbert space of $\mathscr{W}$. We suppose that $\mathscr{H}$ and $\mathscr{K}$ are m-dimensional, while $\mathscr{L}$ is supposed two-dimensional. The notion of a distinguished state in this more general case is similar to that given for two-dimensional spaces.

Definition 4.2 We say that a ket $|\chi\rangle \in \mathscr{M}$ is a distinguished state if the following conditions are satisfied.

(a) The partial density operator ρ_L is generic and with nonzero eigenvalues;
(b) In the canonical representation $|\chi\rangle = d_1 |\psi_1\rangle |l_1\rangle + d_2 |\psi_2\rangle |l_2\rangle$ with $d_1 \neq d_2$, $d_1 > 0$, $d_2 > 0$, $d_1^2 + d_2^2 = 1$, $|l_1\rangle$, $|l_2\rangle$ orthonormal vectors of $\mathscr{L}$ and $|\psi_1\rangle$, $|\psi_2\rangle$ orthonormal vectors of $\mathscr{H} \otimes \mathscr{K}$, $tr_K |\psi_1\rangle \langle\psi_1|$ and $tr_K |\psi_2\rangle \langle\psi_2|$ are generic and have the same eigenstates.

Let $\{|h_i\rangle\}_{i\in\mathbb{N}_m}$ and $\{|k_i\rangle\}_{i\in\mathbb{N}_m}$ be arbitrary ordered orthonormal bases of $\mathscr{H}$ and $\mathscr{K}$, respectively. Let π be a permutation of $\mathbb{N}_m$ which displaces all its points.

Let $\mathbb{N}_m \xrightarrow{\mathbf{m}} \mathbb{R}_\geq$ and $\mathbb{N}_m \xrightarrow{\mathbf{m}'} \mathbb{R}_\geq$ be injective mappings with $\sum m_i = 1$ and $\sum m'_i = 1$. Finally, let $|l_1\rangle$, $|l_2\rangle$ be orthonormal vectors of $\mathscr{L}$ and $d_1 \neq d_2$, $d_1 > 0$, $d_2 > 0$. Put

$$|\psi_1\rangle = \sum \sqrt{m_i}\, |h_i\rangle\, |k_i\rangle, \qquad |\psi_2\rangle = \sum \sqrt{m'_i}\, |h_i\rangle\, \left|k_{\pi(i)}\right\rangle, \tag{4.5.1}$$

and

$$|\chi\rangle = d_1 |\psi_1\rangle |l_1\rangle + d_2 |\psi_2\rangle |l_2\rangle . \tag{4.5.2}$$

We can easily prove the following.

Theorem 4.15 *$|\chi\rangle$ is a distinguished state.*

Proof As $\pi(i) \neq i$, $|\psi_1\rangle$ and $|\psi_2\rangle$ are orthonormal. Hence, (4.5.2) is the canonical representation of $|\chi\rangle$. The corresponding density operator ρ_L is given by

$$\rho_L = d_1^2 |l_1\rangle \langle l_1| + d_2^2 |l_2\rangle \langle l_2| .$$

Therefore, ρ_L is generic and with nonzero eigenvalues. Furthermore

$$tr_K |\psi_1\rangle \langle\psi_1| = \sum m_i\, |h_i\rangle \langle h_i|,$$
$$tr_K |\psi_2\rangle \langle\psi_2| = \sum m'_i\, |h_i\rangle \langle h_i|,$$

so that they are generic and with the same eigenstates. □

Let us evaluate the density operator ρ_H of $|\chi\rangle$ in $\mathscr{R}$. We have

$$|\chi\rangle = d_1 \sum \sqrt{m_i}\, |h_i\rangle |k_i\rangle |l_1\rangle + d_2 \sum \sqrt{m'_i}\, |h_i\rangle \left|k_{\pi(i)}\right\rangle |l_2\rangle , \tag{4.5.3}$$

which we recast in the form $|\chi\rangle = \sum |h_i\rangle |\theta_i\rangle$ with

$$|\theta_i\rangle = d_1\sqrt{m_i}\, |k_i\rangle |l_1\rangle + d_2\sqrt{m'_i}\, \left|k_{\pi(i)}\right\rangle |l_2.\rangle \tag{4.5.4}$$

The kets $|\theta_i\rangle$ are pairwise orthogonal, so that

$$\rho_H = \sum \langle\theta_i|\theta_i\rangle\, |h_i\rangle \langle h_i| = d_1^2 \rho_1 + d_2^2 \rho_2,$$

where ρ_1 and ρ_2 are the density operators in $\mathscr{R}$ corresponding to $|\psi_1\rangle$ and $|\psi_2\rangle$.

We now discuss the probabilities.

Putting $d_1^2 = \lambda$, $d_2^2 = \lambda'$, the spectral decomposition of the density operator ρ_{HK} is $\rho_{HK} = \lambda |\psi_1\rangle \langle\psi_1| + \lambda' |\psi_2\rangle \langle\psi_2|$. Hence, the system $\mathscr{RS}$ is in the state $|\psi_1\rangle$ with probability q_1 and in the state $|\psi_2\rangle$ with probability q_2. Applying the results of previous section, we have $q_1 = \lambda$ and $q_2 = \lambda'$.

Furthermore, the spectral expansion of the density operator of $|\psi_1\rangle$ in $\mathscr{R}$ is $\rho_1 = \sum m_i |h_i\rangle \langle h_i|$, so that if $\mathscr{RS}$ is in $|\psi_1\rangle$, $\mathscr{R}$ is in $|h_i\rangle$ with probability $q_i^{(1)}$. If all the components of $\mathbf{m}$ are nonzero, the rank of ρ_1 is m, so that we have $\mathbf{q}^{(1)} = g_m(\mathbf{m})$.

Similarly, the spectral expansion of the density operator of $|\psi_2\rangle$ in $\mathscr{R}$ is $\rho_2 = \sum m'_i |h_i\rangle \langle h_i|$, so that if $\mathscr{RS}$ is in $|\psi_2\rangle$, $\mathscr{R}$ is in $|h_i\rangle$ with probability $q_i^{(2)}$. If all the components of $\mathbf{m}'$ are nonzero, the rank of ρ_2 is m, so that we have $\mathbf{q}^{(2)} = g_m(\mathbf{m}')$.

But then the probability that $\mathscr{R}$ is in $|h_i\rangle$ is $\bar{q}_i = q_i^{(1)} q_1 + q_i^{(2)} q_2$, provided that ρ_H is generic. But $\rho_H = \lambda\rho_1 + \lambda'\rho_2$ so that $\rho_H = \sum \left(\lambda m_i + \lambda' m'_i\right) |h_i\rangle \langle h_i|$. Hence, for the validity of our argument we require that $\lambda\mathbf{m} + \lambda'\mathbf{m}'$ belongs to the domain of g_m. We have $\bar{\mathbf{q}} = g_m\left(\lambda\mathbf{m} + \lambda'\mathbf{m}'\right)$.

Hence, by substitution, we get

$$g_m\left(\lambda\mathbf{m} + \lambda'\mathbf{m}'\right) = \lambda g_m(\mathbf{m}) + \lambda' g_m\left(\mathbf{m}'\right). \tag{4.5.5}$$

We must be careful about the range of validity of the above equation. In our derivation, we have supposed $\lambda > 0$, $\lambda' > 0$, and $\lambda \neq \lambda'$. Furthermore, $\mathbf{m}$ and $\mathbf{m}'$ must belong to the domain of g_m, as well as $\lambda\mathbf{m} + \lambda'\mathbf{m}'$.

Suppose now that some single component of $\mathbf{m}$, say m_i, is zero. Then the rank of ρ_1 is $m - 1$ and the possible states of $\mathscr{R}$, given that $\mathscr{RS}$ is in $|\psi_1\rangle$, are all the $|h_l\rangle$ with $l \neq i$. The probability that $\mathscr{R}$ is in $|h_l\rangle$ is therefore

$$\bar{q}_l = q_l^{(1)} q_1 + q_l^{(2)} q_2, \tag{4.5.6}$$

for $l \neq i$ and

$$\bar{q}_i = q_i^{(2)} q_2. \tag{4.5.7}$$

We define $\widehat{\mathbf{m}}_i = (m_1, ..., m_{i-1}, m_{i+1}, ..., m_m)$. We extend the definition of g_m (which is defined for all the m_j different from zero) to a domain $\widehat{\Delta}_m$ where at most a single m_k is zero in the following way:

$$\hat{g}_m(\mathbf{m}) = g_m(\mathbf{m}) \qquad \text{for} \qquad \mathbf{m} \in \Delta_m, \tag{4.5.8}$$

$$\hat{g}_{m,k}(\mathbf{m}) = 0, \quad \hat{g}_{m,i}(\mathbf{m}) = g_{m-1,i}\left(\hat{\mathbf{m}}_k\right) \quad i \neq k \quad \text{when} \quad m_k = 0. \tag{4.5.9}$$

We note that $\widehat{\Delta}_m$ is obtained from Δ_m by adjunction of the subset with nonzero different barycentric coordinates of each face of T_m. We now prove the following theorem.

Theorem 4.16 *The mapping $\hat{g}_m$ satisfies the equation*

$$\hat{g}_m\left(\lambda\mathbf{m}+\lambda'\mathbf{m}'\right)=\lambda\hat{g}_m\left(\mathbf{m}\right)+\lambda'\hat{g}_m\left(\mathbf{m}'\right), \tag{4.5.10}$$

provided that all the arguments belong to $\widehat{\Delta}_m$ *and that* λ *and* λ' *belong to* $[0, 1]$ *that they are different from each other and that* $\lambda+\lambda'=1$.

Proof Indeed, for $\lambda=1$ and $\lambda'=1$, we obtain an identity. Otherwise, but when all the arguments belong to Δ_m Eq. (4.5.10) coincides with Eq. (4.5.5).

If $m_i=0$ for a single i and $\mathbf{m}'\in\Delta_m$, the argument of $\hat{g}_m$ in the left side of (4.5.10) belongs to Δ_m, so that the left side of (4.5.10) is the probability distribution of finding $\mathscr{R}$ in the different states $|h\rangle$; the components of the right side of (4.5.10) coincide with the corresponding right sides of (4.5.6) for $l\neq i$ and with the right side of (4.5.7) for $l=i$. Hence, (4.5.10) is satisfied also in this case.

Exchanging the roles of λ, $\mathbf{m}$ and λ', $\mathbf{m}'$, we see that (4.5.10) is satisfied also when $\mathbf{m}\in\Delta_m$ and $m'_i=0$ for a single i. Suppose now that $m_i=0$ for a single i and $m'_k=0$ for a single k with $k\neq i$. In this case, all the states $|h_l\rangle$ $(l\neq i, l\neq k)$ result from both possibilities $|\psi_1\rangle$ and $|\psi_2\rangle$, while the state $|h_i\rangle$ results only from $|\psi_2\rangle$ and the state $|h_k\rangle$ results only from $|\psi_1\rangle$. We then have

$$\bar{q}_l=q_l^{(1)}q_1+q_l^{(2)}q_2 \qquad (l\neq k, l\neq i), \tag{4.5.11}$$

$$\bar{q}_i=q_i^{(2)}q_2, \tag{4.5.12}$$

$$\bar{q}_k=q_k^{(1)}q_1. \tag{4.5.13}$$

As $\lambda\mathbf{m}+\lambda'\mathbf{m}'\in\Delta_m$, the left side of (4.5.10) is the probability distribution of the states $|h\rangle$, while the components of the right side different from i and k coincide with the corresponding right sides of (4.5.11). The components i and k of the right side of (4.5.10) coincide with the right side of (4.5.12) and (4.5.13), respectively. Hence, also in this case (4.5.10) is satisfied.

Finally, suppose $m_i=m'_i=0$ for a single i. In this case, the ith component of both sides of (4.5.10) is zero. Furthermore, the equations for the probabilities are

$$\bar{q}_l=q_l^{(1)}q_1+q_l^{(2)}q_2 \qquad (l\neq i). \tag{4.5.14}$$

The other components of the left side of (4.5.10) are the corresponding left sides of (4.5.14), and the same for the right sides. □

We are now ready to prove the fundamental result of the above analysis.

Theorem 4.17 *The unique mapping* g_m *expressing the probability law of the states is the identity.*

Proof We will prove the theorem by recursion on m. The theorem is true for $m=2$ (see Sect. 4.4).

Suppose that the theorem is true for $m-1$. Let $\mathbf{m}''\in\widehat{\Delta}_m$. If $m''_i=0$ for a single i, by virtue of (4.5.9) we get $\hat{g}_{m,i}\left(\mathbf{m}''\right)=0$ and $\hat{g}_{m,l}\left(\mathbf{m}''\right)=\hat{g}_{m-1,l}\left(\hat{\mathbf{m}''_i}\right)$ for $l\neq i$. But, for the recursion hypothesis, $\hat{g}_{m-1,l}\left(\widehat{\mathbf{m}}''_i\right)=\widehat{\mathbf{m}}''_i$. Hence $\hat{g}_m\left(\mathbf{m}''\right)=\mathbf{m}''$.

Suppose now that $m''_i \neq 0$ for every i. We can always choose a segment containing $\mathbf{m}''$ whose terminal points $\mathbf{m}$ and $\mathbf{m}'$ belong to $\hat{\Delta}_m$ and to two different faces of T_m, opposite to the vertices i and k, respectively, and such that $\mathbf{m}''$ is not the middle point. Then

$$\hat{g}_m\left(\mathbf{m}''\right) = \hat{g}_m\left(\lambda\mathbf{m} + \lambda'\mathbf{m}'\right) = \lambda\hat{g}_m\left(\mathbf{m}\right) + \lambda'\hat{g}_m\left(\mathbf{m}'\right). \tag{4.5.15}$$

But we have already shown that, for such points, $\hat{g}_m\left(\mathbf{m}\right) = \mathbf{m}$ and $\hat{g}_m\left(\mathbf{m}'\right) = \mathbf{m}'$, so that

$$\hat{g}_m\left(\mathbf{m}''\right) = \lambda\mathbf{m} + \lambda'\mathbf{m}' = \mathbf{m}''. \tag{4.5.16}$$

□

Remembering Theorem 4.8, i.e., that $f\left(\rho_S\right) = g_m\left(\delta_\Xi\left(\rho_S\right) \circ \gamma\right) \circ \gamma^{-1}$, we get $f\left(\rho_S\right) = \delta_\Xi\left(\rho_S\right)$, that is, if $\rho_S = \sum p_\alpha \left|\xi_\alpha\right\rangle\left\langle\xi_\alpha\right|$, $f\left(\rho_S\right)\left(\xi_\alpha\right) = p_\alpha$, that is

Theorem 4.18 *If ρ_s is a generic density operator, the possible states of the system, considered as isolated, are its non-null eigenstates, and the probability of finding the system in the state ξ is the correspondent eigenvalue.*

We thus recover the standard statistic interpretation of the density operator.

Note that these results have not been stated as a postulate, but derived from the mathematical formalism, examining in a detailed way each step of the derivation, and introducing and discussing all the necessary mathematical concepts. This was the main aim of the present work.

Appendix A
Categories

We will take[1] for granted the general notion of a category, but some aspects of the theory relevant for our purposes will be evidenced here. We first remark that it is sometimes convenient to describe a category uniquely in terms of morphisms. This is possible because there is a one-to-one correspondence between the objects and their identity morphisms.

From this point of view, the category is regarded as a collection of morphisms. Each morphism can be multiplied on the left by a unique identity, which will be called its *left identity*; similarly, it can be multiplied on the right by a unique identity, which will be called its *right identity*.

If φ is a morphism, and e', e are its left and right identities, respectively, the *source* of φ is the object whose identity is e, and the *target* is the object whose identity is e'. If $X \xrightarrow{\varphi} X'$, in order to specify the source and the target, we write $e'\varphi e$. If $e''\psi e'''$ is a second morphism, we can compose them with ψ on the left if and only if $e' = e'''$ and the result is $e''\psi e' e' \varphi e = e''\psi\varphi e$. If the sources and the targets have been declared, we can omit the identities in our notation.

A *covariant functor* from a category $\mathscr{K}$ to a category $\mathscr{K}'$ is a mapping $\mathscr{F}$ from $\mathscr{K}$ to $\mathscr{K}'$ such that $\mathscr{F}(\psi\varphi) = \mathscr{F}(\psi)\,\mathscr{F}(\varphi)$. A *contravariant functor* from a category $\mathscr{K}$ to a category $\mathscr{K}'$ is a mapping $\mathscr{F}$ from $\mathscr{K}$ to $\mathscr{K}'$ such that $\mathscr{F}(\psi\varphi) = \mathscr{F}(\varphi)\,\mathscr{F}(\psi)$.

A.1 The Categories of Sets

Consider the class $\mathfrak{C}$ of all sets. We can associate to it a category SET in the following way. The objects of SET are all the sets. Given two sets X and Y we define $Hom\,(X, Y)$

[1] The content of this chapter can be found in Campanella's files *The categories of sets: (2/12/2013), Summary on G-set: (27/11/2013).*

M. Campanella et al., *Interpretative Aspects of Quantum Mechanics*, UNIPA Springer Series,
https://doi.org/10.1007/978-3-030-44207-1

as follows. We consider the power sets $\wp(X)$ and $\wp(Y)$. Let $M(X, Y)$ be the set of all mappings F from $\wp(X)$ to $\wp(Y)$ that send $\varnothing$ in $\varnothing$ and singletons in singletons. Let W_X be the subset of $\wp(X)$ whose elements are all the singletons of X and $V_X = W_X \bigcup \{\varnothing\}$. Hence, $\wp(X) \xrightarrow{F} \wp(Y)$ belongs to $M(X, Y)$ if and only if $\varnothing F = \varnothing$ and $W_X F \subseteq W_Y$. As a consequence, $V_X F \subseteq V_Y$.

Furthermore, we introduce the insertion mapping ι_{V_X} of V_X in $\wp(X)$. We have

$$\iota_{V_X} F \iota_{V_Y} = \iota_{V_X} F.$$

We observe that $M(X, Y) = \varnothing$ if and only if $X \neq \varnothing$ and $Y = \varnothing$. If $M(X, Y) \neq \varnothing$, we define in it an equivalence relation through the rule $F \sim F'$ if and only if $\iota_{V_X} F = \iota_{V_X} F'$.

We put $Hom(X, Y) = \varnothing$ if $M(X, Y) = \varnothing$ and $Hom(X, Y) = M(X, Y)/\sim$ if $M(X, Y) \neq \varnothing$.

We now define the composition law for the morphisms. Let $f \in Hom(X, Y)$ and $f' \in Hom(Y, Z)$. Then $M(X, Y)$ and $M(Y, Z)$ are nonempty. Let F, F' be representatives of f, f', respectively. Consider the composition $F'' = FF'$. As F and F' send $\varnothing$ in $\varnothing$ and singletons in singletons, also F'' does, so that $F'' \in M(X, Z)$ and the latter is nonempty. Suppose that $\hat{F}$ and $\hat{F}'$ represent f and f'. Then $\iota_{V_X} F = \iota_{V_X} \hat{F}$ and $\iota_{V_Y} F' = \iota_{V_Y} \hat{F}'$. We have

$$\iota_{V_X} \hat{F}'' = \iota_{V_X} \hat{F} \hat{F}' = \iota_{V_X} \hat{F} \iota_{V_Y} \hat{F}' = \iota_{V_X} F \iota_{V_Y} F' = \iota_{V_X} F''.$$

Thus, the mapping composition law passes to the quotients, defining a composition law for the morphisms. The axioms for the composition of morphisms in a category are automatically satisfied, as they are satisfied for the mappings.

In the category of sets, there is an object A such that for any X, for any $a \in A$, and for any $x \in X$ there is a unique morphism sending a in x. A singleton satisfies this property. It is obvious that only singletons work, so that A is defined up to isomorphisms.

In an arbitrary category, we say that a morphism φ is *cancellable on the left* if the equation $\varphi\xi = \varphi\zeta$ implies $\xi = \zeta$. We say that a morphism φ *is cancellable on the right* if the equation $\xi\varphi = \zeta\varphi$ implies $\xi = \zeta$.

It is easy to see that in the category of sets φ is cancellable on the left if and only if it is injective and that it is cancellable on the right if and only if it is surjective.

The sufficiency is obvious. For the necessity, suppose that φ is not injective and that it is left-cancellable. Then $\varphi = \varphi\zeta$ must imply $\zeta = e$. But $\varphi = \varphi\zeta$ if and only if every equivalence class modulo is stable under ζ. But not all these classes are singletons, so there are non-identical ζ leaving them invariant, and then we get a contradiction.

Suppose now that φ is not surjective and that it is right-cancellable. Then $\varphi = \zeta\varphi$ must imply $\zeta = e$. But $\varphi = \varphi\zeta$ if and only if every element of the image of φ is a fixed point for ζ. But not all the elements of the target of φ are in its image, so there are non-identical ζ for which all the elements of this image are fixed points, and then we get a contradiction.

A *terminal object* $\mathfrak{T}$ in a category $\mathscr{K}$ is an object such that for every object X in $\mathscr{K}$ there is a single morphism from X to $\mathfrak{T}$. Using the terminology involving only morphisms, an identity e is terminal if for every e' there is a unique left multiplier of e which is also a right multiplier of e'.

If e and e'' are terminal, there is a unique φ such that φe and $e''\varphi$ exist, and a unique ψ such that $\psi e''$ and $e\psi$ exist. But then $\varphi\psi$ and $\psi\varphi$ exist; $\varphi\psi$ is a left multiplier of e'' and a right multiplier of e'', e'' enjoys the same property. As e'' is terminal, $\varphi\psi = e''$. Similarly, $\psi\varphi = e$. Therefore, φ and ψ are isomorphisms. We conclude that there is a unique isomorphism between two terminal elements. As an immediate consequence, the set of endomorphisms of a terminal element reduces to identity.

In a similar way, we introduce the notion of i*nitial object*. An initial object $\mathfrak{I}$ in a category $\mathscr{K}$ is an object such that for every object X in $\mathfrak{K}$ there is a single morphism from $\mathfrak{I}$ to X. Using the terminology involving only morphisms, an identity e is initial if for every e' there is a unique right multiplier of e which is also a left multiplier of e'. If e and e'' are initial, there is a unique φ such that φe and $e''\varphi$ exist, and a unique ψ such that $\psi e''$ and $e\psi$ exist. But then $\varphi\psi$ and $\psi\varphi$ exist; $\varphi\psi$ is a left multiplier of e'' and a right multiplier of e'', e'' enjoys the same property. As e'' is initial, $\varphi\psi = e''$. Similarly, $\psi\varphi = e$. Therefore, φ and ψ are isomorphisms. We conclude that there is a unique isomorphism between two initial elements. As an immediate consequence, the set of endomorphisms of an initial element reduces to identity.

In the category of sets, an object is terminal if and only if it is a singleton. Indeed, in SET the only sets having the identity as their unique endomorphism are the singletons. On the other hand, the only mapping toward a singleton is the constant map.

On the contrary, there is no initial object in SET. For if there were one, it ought to be a singleton, but there is a bijection between the elements of X and the morphisms from a singleton to X. However, it is just the latter property that allows us to describe the elements of X in terms of morphisms. Once we fix a particular singleton, we can select any element of X through the corresponding morphism, which for this reason is called a selection morphism.

Let $\mathscr{K}$ be a concrete category. Let us investigate the question of the existence of terminal and initial objects. We know that for such objects the set of endomorphisms must reduce to identity. If there is some $\Omega \in \mathscr{K}$ whose underlying set is a singleton and such that for every X there is a morphism from $\mathscr{K}$ to Ω, the latter is obviously a terminal object. Of course all terminal objects are isomorphic. However, in many concrete categories such objects are also initial objects, so that the morphisms starting from such an object select just one element for each object of the category, so that the situation is quite trivial. This is what happens, for instance, in the category of linear spaces or in that of groups. This problem is present in every category where a terminal object is a singleton and is also an initial object.

We can overcome this trivial situation in some concrete categories in the following way. Starting from a concrete category $\mathscr{K}$, we define a category $\mathscr{K}_\Theta$ whose morphisms are all the morphisms of $\mathscr{K}$ and in addition all the mappings from a selected terminal element Θ of SET (that is a singleton) to all the objects of $\mathscr{K}$ and define the morphisms of a new category $\Theta\mathscr{K}$ as follows. Let F_Θ be the set of all the mappings from Θ to all the objects of $\mathscr{K}$. For $x \in F_\Theta$, we denote e_x the left unit of x. The morphisms of $\Theta\mathscr{K}$ are the words of the form $e_g \tau e_f$ such that $g = \tau f$. Two words can be composed if and only if the right symbol of the left word is the same as the left symbol of the right word and the outcome of the composition is obtained suppressing the intermediate units and composing the mappings. It is immediate to check that what we have just defined is indeed a category. If we change Θ in Θ', we can define a covariant functor $\mathscr{F}$ from $\Theta\mathscr{K}$ to $\Theta'\mathscr{K}$ putting $\mathscr{F}\left(e_g \tau e_f\right) = e_{g\iota} \tau e_{f\iota}$ where ι is the mapping from Θ' to Θ. But e_x depends only on the image of x. We conclude that $\Theta\mathscr{K}$ is independent of Θ. We will denote $\mathscr{K}^\frown$ this category. It can happen when $\mathscr{K}^\frown$ has an initial element. In this case, it is unique up to isomorphisms. Let $\mathfrak{I} = \Theta_0 \xrightarrow{\vartheta_0} X_0$ an initial element. Therefore, for every $X \in \mathscr{K}$ and every $\{x\} \subseteq X$, there is a unique $\mathscr{K}$-morphism $X_0 \xrightarrow{\tau} X$ such that $\tau \circ \vartheta_0\left(\Theta_0\right) = \{x\}$. With an abuse of terminology, we say that X_0 is an initial element of $\mathscr{K}$ if X_0 is the target of some initial element of $\mathscr{K}^\frown$.

We can show that:

If $(S,\mathfrak{M})$ is a set S equipped with a monoid $\mathfrak{M}$ of mappings of S into itself, it is an initial element of some category if and only if there is $x_0 \in S$ such that for every $x \in S$ there is a unique $m \in \mathfrak{M}$ such that $x = mx_0$.

Indeed the necessity follows if we take $X = S$. For the sufficiency, we observe that $(S,\mathfrak{M})$ itself is a category with $(S,\mathfrak{M})$ as its initial element.

Let us investigate the structure of an initial element. If x_0 is the image of an initial element of $\mathscr{K}^\frown$, for every $x \in S$ there is a unique $m \in \mathfrak{M}$ such that $x = mx_0$, so that a bijective mapping $S \xrightarrow{\varphi_{x_0}} \mathfrak{M}$ arises. The orbit of x_0 under the action of $\mathscr{G}$ must be simply transitive. As an initial element of $\mathscr{K}^\frown$ is defined up to isomorphisms, we can replace S with $\mathfrak{M}$ and Θ_0 with the identity of $\mathfrak{M}$. Consistently, x_0 will be some element m_0 of $\mathfrak{M}$. The mapping ϑ_0 sends from the identity id of $\mathfrak{M}$ to m_0. The mapping φ_{x_0} becomes a bijective mapping μ_{m_0} from the set $\mathfrak{M}$ to the monoid $\mathfrak{M}$. If x belongs to the set $\mathfrak{M}$, $\mu_{m_0}(x)$ is the unique m such that $x = mm_0$. There is a bijection between the mappings ϑ_0 and the elements m_0. As $\mu_{m_0}(id) = m_0$, the mapping which associates each image m_0 of ϑ_0 with μ_{m_0} is injective. We denote $\mathscr{G}$ the group of the invertible elements of $\mathfrak{M}$. Let ϑ_0 and ϑ'_0 define initial elements and let m_0 and m'_0 be the corresponding images. Then there are unique m' and m'' such that $m'_0 = m'm_0$ and $m_0 = m''m'_0$. We get $m_0 = m''m'm_0$ and $m'_0 = m'm''m'_0$, so that $m''m' = m'm'' = id$. Therefore $m' = g \in \mathscr{G}$ and $m'' = g^{-1}$. Defining a left action of $\mathscr{G}$ on $\mathfrak{M}$ by left multiplication, we conclude that, if m_0 is the image of an initial element, every image of an initial element lies on the orbit of m_0 under this action. If we take $m_0 = id$, we conclude that every image of an initial element belongs to $\mathscr{G}$. Conversely, every element of $\mathscr{G}$ is the image of an initial element. In fact, the equation $m = xg$ has the unique solution $x = mg^{-1}$. We conclude that the set of images of the initial elements is the group of the invertible elements of $\mathfrak{M}$.

A.2 The Category of G-Sets

We recall here some useful notions and properties concerning G-sets. We recall that, for a given group G, a G-set is a set X equipped with a group morphism from G in the group of permutations of X.

For technical reasons to be seen later, the notion of a G-set will be needed also when X empty. In this case, our definition is clear only once we clarify what is to be understood as the group of permutations of the empty set. This can be easily explained if we regard sets as the objects of the category SET, and mappings between nonempty sets as the morphisms of this category. Now, as long as X is nonempty, the permutations of X are the bijective mappings of X into itself, that is, the automorphisms of X as an object of SET. The notation $Hom\ (\diamond)$ will be employed to denote the sets of morphisms between objects of SET, and $End\ (\diamond)$ $Aut\ (\diamond)$ to denote the sets of endomorphisms, automorphisms of the objects of SET. For any other category, we will write hom $(\diamond)$, $end\ (\diamond)$, and $aut\ (\diamond)$. Thus, we can slightly modify our definition in order to include also the case of $X = \varnothing$:

Definition A.1 For a given group G, a G-set is a set X equipped with a group morphism ϑ from G in the group of automorphisms of X. Thus, for $X = \varnothing$ $Aut\ (X) = \left\{id_{\varnothing}\right\}$ and the unique morphism of G in $Aut\ (\varnothing)$ is trivial.

For future convenience, the value of a map f in x will be denoted xf rather than with the more usual "functional" notation $f\ (x)$. Consistently, the composition of mappings, and more generally, of morphisms in a category will be expressed in the reverse order with respect to the usual one.

The homomorphism ϑ is called the *action* of G on X. Thus, an action is characterized by the condition

$$\left(gg'\right)\vartheta = (g\vartheta)\left(g'\vartheta\right).$$

Hence, a G-set is characterized by a pair (X, ϑ).

Let X, Y be G-sets and $X \stackrel{\gamma}{\to} Y$ be an element of $Hom\ (X, Y)$. We say that γ is a *morphism* of G-sets if

$$(g\vartheta)\,\gamma = \gamma\left(g\vartheta'\right).$$

We will denote hom (X, Y) the set of all morphisms of G-sets from X to Y. The G-sets together with their morphism form a full subcategory of SET. This category will be denoted as G-SET.

Let G be a group and H a subgroup of G. Let G/H be the set of the *right cosets* of H in G, i.e., *the set of subsets of* G *of the form* $R = Hg$. The position $R \mapsto Rg$ for every $R \in G/H$ defines an action of G on G/H. In this way, G/H becomes a G-set. The terminology adopted here emphasizes the side of the action of G on G/H rather than the side of H in the definition of the cosets. We observe that any $R = Hg$ can be expressed as $(He)\,g$, so that the coset He can be brought by the action of G on any coset, and hence this action is able to bring any coset on any other. We say that the action is *transitive*. Furthermore, the set of solutions of the equation $(He)\,g = He$ is the subgroup H.

We can multiply both members of an equality $x = x'$ on the right by an arbitrary g, obtaining the equality $xg = x'g$. This happens because this condition is a shorthand for $x\,(g\vartheta) = x'\,(g\vartheta)$.

A.2.1 Congruence. Invariant Sets

Consider now an equivalence relation $\equiv$ on X which "behaves as identity," that is such that $\left(x \equiv x'\right) \Rightarrow xg \equiv x'g$. Such an equivalence relation is called a *congruence* and the corresponding classes of equivalence are called *classes of congruence*. We denote C_x the unique class of congruence containing $\{x\}$.

If $\sim$ is a congruence, the image Cg of a class of congruence C is a class of congruence.

Indeed, if $x \in C$ we have

$$Cg = \left\{x'g | x' \sim x\right\} = \left\{x'g | x'g \sim xg\right\},$$

so that Cg is the class of congruence of xg. Incidentally, we have also $C_{xg} = C_x g$.

We say that a mapping $X \xrightarrow{\gamma} Y$ is a *morphism* of G-sets if $(x\,(g\vartheta))\,\gamma = (x\gamma)\left(g\vartheta'\right)$. On the other hand, $(x\,(g\vartheta))\,\gamma = x\,((g\vartheta)\,\gamma) = x\,(g\vartheta)\,\gamma$. If $xg\gamma$ is a shorthand notation for $x\,(g\vartheta)\,\gamma$ we can write $(x\,(g\vartheta))\,\gamma = xg\gamma$. Furthermore $(x\gamma)\left(g\vartheta'\right) = x\left(\gamma\left(g\vartheta'\right)\right) = x\gamma\left(g\vartheta'\right)$, so that, if $x\gamma g$ is a shorthand notation for $x\gamma\left(g\vartheta'\right)$, we get $xg\gamma = x\gamma g$.

Both sides of this equation define mappings, so that $g\gamma = \gamma g$. Hence, in this notation, *the condition for γ being a morphism is that it must commute with the action*. However, we emphasize again that the mapping $g\gamma$ is an abbreviation of $(g\vartheta)\,\gamma$, while the mapping γg is an abbreviation of $\gamma\left(g\vartheta'\right)$.

We recall that in a general category (in our reversed notation), *a morphism which is right-cancellable is called a mono, a morphism which is left-cancellable is called an epi, and a morphism which is both mono and epi is called an isomorphism*. Equivalently, an isomorphism is a morphism which admits a two-sided inverse, which is unique.

In the category of sets, a morphism is a mapping and it is a mono (an epi, an iso) if and only if it is injective (surjective, bijective). In a category whose objects are sets and whose morphisms are mappings, both notions of mono and injective (epi and surjective, iso and bijective) make sense, but they must hold distinct in general. However, *there are categories whose objects are sets and whose morphisms are mappings, for which these notions are pairwise equivalent, as it happens in* SET. *It can be shown that this is the case for the category of G-sets.*

If $X \xrightarrow{f} Y$ is a mapping, we define the *kernel of* f *as the equivalence relation* $\sim_f$ *specified by the rule* $x \sim_f x'$ *if and only if* $xf = x'f$. Sometimes we will understand the kernel of f as the set of the equivalence classes of $\sim_f$.

An equivalence relation $\sim$ *on a* G-set X *is a congruence if and only if it is the kernel of some morphism.*

Indeed, suppose that $\sim = \ker(\gamma)$. If $x \sim x'$, $x\gamma = x'\gamma$ and $xg\gamma = x\gamma g = x'\gamma g = x'g\gamma$, so that $xg \sim x'g$. Conversely, suppose that $\sim$ is a congruence. Define the canonical projection $X \xrightarrow{\pi} X/\!\sim$. This mapping is defined through the position $x\pi = [x]$, where $[x]$ is the class of congruence of x, so that $(x\pi = x'\pi) \Leftrightarrow x \sim x'$ and then $\sim = \ker(\pi)$.

We define on $X/\!\sim$ an action $*$ through the position $[x] * g = [xg]$. This definition is well posed. Indeed, if $[x] = [x']$, $x' \sim x$ and $x'g \sim xg$, so that $[x'g] = [xg]$. Under this action on $X/\!\sim$ π is a morphism. Indeed, $xg\pi = [xg] = [x] * g = x\pi g$. Hence, $\sim$ is the kernel of a morphism.

We observe that, defining for any subset S of X the set $Sg = \{sg | s \in S\}$, we have $[xg] = [x]g$. Indeed, if $x' \in [xg]$, $x' \sim xg$. But

$$[x]g = \{x''g | x'' \sim x\} = \{x''g | x''g \sim xg\} = \{w | w \sim xg\},$$

so that $x' \in [x]g$. Conversely, let $x' \in [x]g$. Then $x' = wg$ for some $w \sim x$. Hence, $x' \sim xg$, so that $x' \in [xg]$. As a consequence, $[x] * g = [x]g$. This equation means that the action $*$ on the elements of the set $X/\!\sim$ is the same as the natural action induced on these elements as subsets of X. We will use the notation $[x]g$ to understand both interpretations.

Given a congruence $\sim$ on X we can consider the *category* $\mathscr{C}_\sim$ *whose objects are all the morphisms whose kernel is* $\sim$. Given two objects γ and γ', we define hom (γ, γ') *as the set of all morphisms* α *such that* $\gamma' = \gamma\alpha$. We can easily show that the *canonical projection* π *on* $X/\!\sim$ *is an initial object of* $\mathscr{C}_\sim$.

Indeed, let $\gamma \in Ob\, \mathscr{C}_\sim$. As $\ker(\gamma) = \sim$, we have the canonical decomposition $\gamma = \pi\iota$ as mappings, where ι is injective. As π is surjective, each $y \in X/\!\sim$ can be expressed as $x\pi$. Hence

$$yg\iota = x\pi g\iota = xg\pi\iota = xg\gamma = x\gamma g = x\pi\iota g = y\iota g.$$

Hence, the above canonical decomposition can be regarded as a morphism decomposition. If $\gamma = \pi\iota'$, we have $\pi\iota' = \pi\iota$. As a mapping, π is surjective and hence left-cancellable. We thus conclude that $\iota = \iota'$ and this condition can be interpreted as an equality of morphisms, whence the uniqueness of the decomposition. This shows that π is an initial object.

It is well known that, if a category admits an initial object, for any pair of them there is a unique isomorphism between them. In our case, an arbitrary initial object has the form πj where j is an arbitrary isomorphism whose domain is $X/\sim$. Hence, an object of $\mathscr{C}_\sim$ is an initial object if and only if it is an epi.

For every pair of epis ε and ε' whose domain is X, we put $\varepsilon' \sim \varepsilon$ if and only if there is an isomorphism j such that $\varepsilon' = \varepsilon j$. It is clear that this position defines an equivalence relation in the class of all epimorphisms whose domain is X. ε and ε' are equivalent if and only if they have the same kernel. Indeed, if $\varepsilon' = \varepsilon j$, $x\varepsilon = x'\varepsilon \Leftrightarrow x\varepsilon' = x'\varepsilon'$.

Conversely, if $\ker(\varepsilon) = \ker\left(\varepsilon'\right) = \sim$, ε and ε' are initial objects of the category $\mathscr{C}_\sim$, so that $\varepsilon' = \varepsilon j$.

Henceforth, a class of equivalence of epis from X will be called a *quotient object* (shortly *quobject*) of X. Thus, a mapping arises which associates to each quobject of X the congruence of X given by the kernel of any member of the quobject.

If $\sim$ is a congruence of X, we can associate to it the equivalence class of its projection on $X/\sim$, giving rise to a mapping from the set of congruences of X to the set of the quobjects of X. The two mappings just defined are mutually inverse, so that they are bijective.

If $X \xrightarrow{\gamma} Y$ is a morphism, we can associate to it the quobject of X corresponding to $\ker\gamma \triangleq \sim$. An epi ε belongs to this quobject if and only if $\gamma = \varepsilon\mu$, where μ is a mono. Indeed, γ has the canonical decomposition $\gamma = \pi\tilde{\mu}$ with $\ker\pi = \sim$ and where $\tilde{\mu}$ is a mono. Thus, ε belongs to the class corresponding to $\sim$ if and only if $\pi = \varepsilon j$ where j is an isomorphism. Hence, $\gamma = \varepsilon j\tilde{\mu} = \varepsilon\mu$ where $\mu = \tilde{\mu} j$ is a mono.

For an arbitrary G-set and for arbitrary subsets $S \subseteq X$ and $R \subseteq G$, we define SR as the set $\{sr | r \in R, s \in S\}$. If $S = \{x\}$ $(R = \{g\})$ we will use the notation

$$xR \triangleq \{x\}R \quad (Sg \triangleq S\{g\}).$$

In particular, if R is a subgroup of G, any $s \in S$ can be expressed as $s = se$ with $s \in S$, $e \in R$, so that $S \subseteq SR$.

We say that a subset $J \subseteq X$ is an invariant set if and only if $JG \subseteq J$. Thus, a subset $J \subseteq X$ is an invariant set if and only if $JG = J$.

We can easily show that a subset J of a G-set X is an invariant set if and only if it is the image of a morphism. Indeed, if $J = \mathrm{Im}(\gamma)$, and $y \in J$, $y = x\gamma$ for some $x \in X$, so that $yg = x\gamma g = xg\gamma \in J$. Conversely, let J be an invariant set. Hence, for any $x \in J$ and any $g \in G$ $xg \in J$, so that an action of G on J arises. This action will be called the *induced action* on J. Equipping J with the induced action, the insertion ι of J in X is a morphism. Indeed, $xg\iota = xg = x\iota g$. Furthermore, $J = \mathrm{Im}(\iota)$.

Given an invariant set J of X we can consider the category $\mathscr{C}_J$ whose objects are all the morphisms into X whose image is J. Given two objects γ and γ', we define $\hom\left(\gamma', \gamma\right)$ as the set of all morphisms α such that $\gamma' = \alpha\gamma$. We can easily show that *the insertion ι of J is a terminal object of $\mathscr{C}_J$*. Indeed, let $\gamma \in Ob\,\mathscr{C}_J$. As $\mathrm{Im}(\gamma) = J$, we have the canonical decomposition $\gamma = \varepsilon\iota$ as mappings, where ε is surjective. We have

$$xg\varepsilon\iota = xg\gamma = x\gamma g = x\varepsilon\iota g = x\varepsilon g\iota.$$

Hence $g\varepsilon\iota = \varepsilon g\iota$. As ι is injective, it is right-cancellable, so that ε is a morphism. Hence, the above canonical decomposition can be regarded as a morphism decomposition. If $\gamma = \varepsilon'\iota$, we have $\varepsilon'\iota = \varepsilon\iota$ As a mapping, ι is injective and hence right-cancellable. We thus conclude that $\varepsilon = \varepsilon'$ and this condition can be interpreted as an equality of morphisms, whence the uniqueness of the decomposition. We conclude that ι is a terminal object.

It is well known that, if a category admits a terminal object, for any pair of them there is a unique isomorphism between them. In our case, an arbitrary terminal object has the form $j\iota$ where j is an arbitrary isomorphism whose image is J. Hence, an object of $\mathscr{C}_J$ is a terminal object if and only if it is a mono.

For every pair of monos μ and μ' whose codomain is X, we put $\mu' \sim \mu$ if and only if there is an isomorphism j such that $\mu' = j\mu$. It is clear that this position defines an equivalence relation in the class of all monomorphisms whose codomain is X. μ and μ' are equivalent if and only if they have the same image. Indeed, suppose that $\mu' = j\mu$ if $x \in \operatorname{Im}\mu, x = y\mu = yj^{-1}\mu'$, so that $\operatorname{Im}\mu \subseteq \operatorname{Im}\mu'$. Exchanging the roles of μ and μ' we obtain the reverse inclusion, so that $\operatorname{Im}\mu = \operatorname{Im}\mu'$. Conversely, let $\operatorname{Im}\mu = \operatorname{Im}\mu' = J$. Then μ and μ' are terminal objects of the category $\mathscr{C}_J$, so that $\mu' = j\mu$.

Henceforth, a class of equivalence of monos into X will be called a *subobject* of X. Thus, a mapping arises which associates to each subobject of X the invariant subset of X given by the image of any member of the subobject.

If J is an invariant set of X, we can associate to it the equivalence class of its insertion in X, giving rise to a mapping from the set of invariant subsets of X to the set of the subobjects of X. The two mappings just defined are mutually inverse, so that they are bijective.

If $Y \xrightarrow{\gamma} X$ is a morphism, we can associate to it the subobject of X corresponding to $\operatorname{Im}\gamma \triangleq J$. A mono μ belongs to this subobject if and only if $\gamma = \varepsilon\mu$, where ε is an epi. Indeed, γ has the canonical decomposition $\gamma = \tilde{\varepsilon}\iota$ with $\operatorname{Im}\iota = J$ where $\tilde{\varepsilon}$ is a mono. Thus, μ belongs to the class corresponding to J if and only if $\iota = j\mu$ where j is an isomorphism. Thus, $\gamma = \tilde{\varepsilon}j\mu = \varepsilon\mu$ where $\varepsilon = \tilde{\varepsilon}j$ is an epi.

We can summarize the above discussion as follows:

There is a bijection between the quobjects of a G-set X and the congruences of X. The corresponding of each quobject is the kernel of any member of it. The corresponding of each congruence is the quobject containing the canonical projection of the congruence. If $X \xrightarrow{\gamma} Y$ is a morphism, γ admits factorizations of the form $\gamma = \varepsilon\mu$, where ε is epi and μ is mono. In any such factorization, the left factor is a representative of the quobject corresponding to the kernel of γ.

There is a bijection between the subobjects of a G-set X and the invariant subsets of X. The corresponding of each subobject is the image of any member of it. The corresponding of each invariant subset is the subobject containing the canonical *insertion of the invariant subset. If* $Y \xrightarrow{\gamma} X$ *is a morphism,* γ *admits factorizations of the form* $\gamma = \varepsilon\mu$, *where* ε *is epi and* μ *is mono. In any such factorization, the right factor is a representative of the subobject corresponding to the image of* γ.

In this way, we see that *the notions of a congruence and of an invariant set can be formulated in pure category theoretical terms and they are dual in this sense.*

Let X *be a* G*-set and* Π *a partition of the set* X *consisting of invariant subsets of the* G*-set* X. *Then the equivalence relation* $\sim$ *defined by* Π *is a congruence.* Indeed, let $x \sim x'$. Then there is a unique $J \in \Pi$ such that $x \in J, x' \in J$. As J is invariant, $xg \in J, x'g \in J$, so that $xg \sim x'g$. If J is an invariant set of X, its complement $\bar{J}$ in X is an invariant set. Indeed, let $x \in \bar{J}$. Suppose that $xg \notin \bar{J}$. Hence $xg \in J$, so that $x = (xg)\, g^{-1} \in J$, a contradiction.

Let $J \subseteq X$ be an invariant set. Then the partition $X = J \bigcup \bar{J} \triangleq \Pi_J$ consists of invariant sets. Thus, the corresponding equivalence relation is a congruence. In this way, a mapping arises which associates to each invariant subset a congruence. This congruence defines a corresponding quobject of X. Hence, we get a mapping Υ from the set of subobjects of X to the set of its quobjects. Let us find the image Im Υ of this mapping. If $\varepsilon \in \text{Im}\ \Upsilon$, then there is an invariant subset $J \subseteq X$ such that ε is the quobject corresponding to the congruence defined by the partition Π_J.

A representative of this quobject is the canonical projection on the quotient G-set modulo the congruence. This G-set is the partition $\Pi_J \triangleq \{J, \bar{J}\}$ equipped with the trivial action of G on it. The corresponding canonical projection is

$$\pi_J : x\pi_J = J(x \in J);\ x\pi_J = J(x \in \bar{J}).$$

Thus, $J\Upsilon$ is an equivalence class of epimorphisms from X to a trivial G-set with two nontrivial subobjects. We have $x \sim x'$ if and only if either they both belong to J, or they both belong to $\bar{J}$.

Let $X \xrightarrow{\gamma} Y$ *be a morphism. There is a unique decomposition* $\gamma = \pi j$ *in where* π *is a projection morphism on a quotient* G-set, *in is an insertion morphism and* j *is an isomorphism.*

Indeed, considering γ as a mapping, there is a unique such decomposition as a composition of mappings. In this decomposition, π projects on $X/\ker(\gamma)$ and *in* inserts Im (γ) in Y. As ker (γ) is a congruence and Im (γ) is an invariant set, π and *in* are morphisms. It remains to prove that the bijection j is a morphism. We have $g\gamma = g\pi j\ in = \pi g j\ in = \gamma g = \pi j\ in g = \pi j g\ in$. Thus $\pi g j\ in = \pi j g\ in$. As π is left-cancellable and *in* is right-cancellable, $gj = jg$ and j is a morphism. Hence, the decomposition in mappings is also a decomposition in morphisms. The uniqueness follows from the fact that a decomposition in morphisms is also a decomposition in mappings, and the latter is unique.

Given a set S, a *free* G*-set* over S is a mapping $S \xrightarrow{\iota} F(S)$ into the G-set $F(S)$ such that for every mapping $S \xrightarrow{l} X$ there is a unique morphism $F(S) \xrightarrow{\gamma} X$ of G-sets such that $l = \iota\gamma$.

Let us consider the set $F = S \times G$. For every $z = (s, \bar{g}) \in F$ and every $g \in G$, we define $zg = (s, \bar{g}g)$. In this way, F is equipped with a structure of G-set. We further define $S \xrightarrow{\iota} F : s\iota = (s, e)$. For each $S \xrightarrow{l} X$ we put $z\gamma = (s, g)\,\gamma = (sl)\,g$. We have $s\iota\gamma = sl$. The mapping γ is a morphism. Indeed, $zg\gamma = (s, \bar{g})\,g\gamma = sl \cdot \bar{g}g = z\gamma g$. For an arbitrary $z \in F$ we can write $z = (s, e)\,g$. Let γ' be such that $l = \iota\gamma'$. Hence $(s, e)\,\gamma' = (s, e)\,\gamma$. Consequently

$$z\gamma' = (s, e)\,g\gamma' = (s, e)\,\gamma' g = (s, e)\,\gamma g = (s, e)\,g\gamma = z\gamma$$

which proves the uniqueness of γ. We conclude that $F(S) = S \times G$ with the action $(s, \bar{g})\,g = (s, \bar{g}g)$ and with the mapping $S \xrightarrow{\iota} S \times G : s\iota = (s, e)$ is a free G-set over S. According to a general property of a free object in a category, if $S \xrightarrow{\iota'} F'(S)$ is an arbitrary free G-set over S, there is a unique isomorphism $S \times G \xrightarrow{\eta} F'(S)$ such that $\iota' = \iota\eta$. We will call the free G-set $S \xrightarrow{\iota} S \times G : s\iota = (s, e)$ the *standard G-set over S*.

The mapping ι is clearly injective. Furthermore, $S\iota$ generates $S \times G$. Indeed, any invariant set containing $S\iota$ must contain all the elements of the form $(s, e)\,g$ together with (s, e). Every element of $S \times G$ can be expressed as $(s, e)\,g$, so that $J(S\iota) = S \times G$. We observe that the representation of (s, g) as $(s, e)\,g$ is unique, because

$$(s, e)\,g = \left(s', e\right)g' \Rightarrow (s, g) = \left(s', g'\right).$$

A.2.2 Orbits. Stabilizer

If X is a G-set, we introduce in it a relation $\tilde{\Omega}$ by declaring $x\tilde{\Omega}x'$ if and only if there is $g \in G$ such that $x' = xg$. $\tilde{\Omega}$ is clearly an equivalence relation.

An *orbit* in X is a class of equivalence of $\tilde{\Omega}$. Really, $\tilde{\Omega}$ is a congruence. Indeed, if $x\tilde{\Omega}x'$, $x' = x\bar{g}$ for some $\bar{g}$, so that $x'g = x\bar{g}g = xg\left(g^{-1}\bar{g}g\right)$ and then $xg\tilde{\Omega}x'g$. The congruence $\tilde{\Omega}$ will be called *orbit congruence*, and the set of its classes of congruence (the orbits) will be denoted as Ω.

Clearly, if J is an invariant set of X, considering it as a G-set for the induced action, the orbit congruence for J is the restriction to it of the orbit congruence for X. Consequently, an orbit of the G-set J is an orbit of the G-set X and an orbit of the G-set X contained in J is also an orbit for the G-set J.

We say that a G-set is *homogeneous* if it consists of a single orbit. We say that an invariant set is homogeneous if it is a homogeneous G-set for the induced action.

If X is homogeneous, the only nonempty G-invariant subset of X is X itself. Indeed, if there were a nontrivial G-invariant subset J in X, choosing $x' \notin J$, we would have $J \subset J' \triangleq J \bigcup \left\{x'\right\}$. As X is homogeneous, choosing $x \in J$, there would be $g \in G$ such that $x' = xg$. But this is in contradiction with the invariance of J, because g brings from an element of J to an element outside J.

A nonempty invariant set J is called *minimal* if for every nonempty invariant set J', $J' \subseteq J \Rightarrow J' = J$. For a nonempty invariant set, the following propositions are equivalent:

$$1) J \text{ is minimal,}$$
$$2) J \text{ is an orbit of } X,$$
$$3) J \text{ is homogeneous.}$$

2) $\Leftrightarrow$ 3). Indeed, J is an orbit of X and is contained in J, so that it is also an orbit of itself, and therefore J consists of a single orbit for the induced action, and then it is homogeneous. Conversely, if it is homogeneous, it consists of a single orbit. This orbit is also an orbit of X.
1) $\Rightarrow$ 2). Let J be a minimal invariant set. Its decomposition in orbits must consist of a single orbit, otherwise J would not be minimal. This single orbit is also an orbit of X.
3) $\Rightarrow$ 1). As J is homogeneous, its only nonempty invariant set is J itself, so that J is minimal.

Given an element x of a G-set X, the *stabilizer* of x is the subgroup

$$H_x \triangleq \{g \in G | xg = x\}.$$

It is immediate to show that $H_{xg} = g^{-1} H_x g$.

Two elements $x, x' \in X$ are called *costable* if and only if $H_x = H_{x'}$. Clearly the costability is an equivalence relation. A costability class is a class of equivalence of the costability relation. Really, costability is a congruence. Indeed, if x and x' are costable, then $H_x = H_{x'}$. We have $H_{xg} = g^{-1} H_x g = g^{-1} H_{x'} g = H_{x'g}$, so that xg and $x'g$ are costable.

Let $X \xrightarrow{\gamma} Y$ be a morphism. We have

$$x = xg \;\Rightarrow\; x\gamma = xg\gamma \;\Leftrightarrow\; x\gamma = x\gamma g \;\Leftrightarrow\; g \in H_{g\gamma}.$$

Hence $H_x \subseteq H_{x\gamma}$. If γ is an isomorphism, $H_x = H_{x\gamma}$.

A homogeneous G-set is called a *torsor* if the stabilizer of a (and hence of every) point is $\{e\}$. It is clear that, if X is a torsor, the equation $y' = yg$ has a unique solution for g. For future convenience, we introduce a special notation for this solution. Namely, we put $yy'\theta \triangleq g$. The following calculation rules are obvious:

$$1) yy\theta = e,$$
$$2) y'y\theta = \left(yy'\theta\right)^{-1},$$
$$3) \left(yy'\theta\right)\left(y'y''\theta\right) = \left(yy''\theta\right).$$

Furthermore, $y\left(y'g\right)\theta = yy'\theta g$. Indeed, if $yy'\theta = g_1$, then $y' = yg_1$ and if $y\left(y'g\right)\theta = g_2, y'g = yg_2$, so that $yg_1g = g_2$, i.e., $y\left(y'g\right)\theta = yy'\theta g$. We also have

$$(yg)\,y'\theta = \left(y'\,(yg)\,\theta\right)^{-1} = g^{-1}yy'\theta.$$

For a morphism γ of G-sets, if ω is an orbit of X, $\omega\gamma$ is an orbit of Y. Indeed $\omega\gamma$ is an invariant subset of Y. Let $y, y' \in \omega\gamma$. Then $y = x\gamma$ for some $x \in \omega$ and $y' = x'\gamma$ for some $x' \in \omega$. As ω is homogeneous, $x' = xg$ for some $g \in G$, so that $y' = xg\gamma = yg$. Hence, $\omega\gamma$ is a homogeneous invariant set, that is, an orbit.

If Y is a homogeneous set, γ is surjective, because, if ω is an orbit of X, $\omega\gamma$ is an orbit, and hence $\omega\gamma = Y$. Furthermore, the image of every orbit is Y.

Let X be a homogeneous G-set. Given $x_0 \in X$, every $x \in X$ can be expressed as $x = x_0g$ for some $g \in G$. We put $x\gamma_{x_0} = H_{x_0}g$. The definition is well posed because if $x = x_0g'$, $g' = hg$ with $h \in H_{x_0}$ so that $H_{x_0}g' = H_{x_0}hg = H_{x_0}g$. In this way, we have defined a mapping $X \xrightarrow{\gamma_{x_0}} G/H_{x_0}$. This mapping is a G-morphism. Indeed, $x\bar{g} = x_0g\bar{g}$, so that $x\bar{g}\gamma_{x_0} = H_{x_0}g\bar{g} = x\gamma_{x_0}\bar{g}$. As G/H_{x_0} is homogeneous, γ_{x_0} is surjective. Suppose that $x\gamma_{x_0} = x'\gamma_{x_0}$. If $x = x_0g$ and $x' = x_0g'$, we have $H_{x_0}g = H_{x_0}g'$, so that $g' = hg$ with $h \in H_{x_0}$ and $x' = x_0hg = x$. We conclude that γ_{x_0} is an isomorphism. Hence, once a point x_0 has been selected in X, a canonical isomorphism is established between X and G/H .

Two homogeneous G-sets are *isomorphic* if and only if they have the same conjugation class of stabilizers. The condition is necessary. Indeed, let $X \xrightarrow{\gamma} Y$ be an isomorphism. For any $y \in YH_y = H_x$, where $x = y\gamma^{-1}$. Hence, the conjugation classes of the stabilizers of Y and X have a common element, so that they coincide. Conversely, suppose that X and Y have the same conjugation class of stabilizers K. Let us choose $x_0 \in X$ and put $H = H_{x_0} \in K$. Hence we have the isomorphism $X \xrightarrow{\gamma_{x_0}} G/H$. Choose now in Y an element y_0 such that $H_{y_0} = H$. As before, we can build an isomorphism $Y \xrightarrow{\gamma_{y_0}} G/H$. We conclude that $\gamma_{x_0}\gamma_{y_0}^{-1}$ is an isomorphism between X and Y.

We say that a subset R (L) of G is a *right (left) coset of G if there is* $H \in Sub\,(G)$ *such that* $Rc(L)$ *is a right (left) coset of* H *in* G.

If R is a right coset of G, H is uniquely determined by R. Indeed, if $R \in G/H$, $R = Hg$ for some $g \in G$. If $R = H'g'$, we have $H' = Hgg'^{-1}$ and also $H = H'\left(gg'^{-1}\right)^{-1}$. Thus $\left(gg'^{-1}\right)^{-1} \in H$, so that $gg'^{-1} \in H$ and $H' = H$. In this way, if $Sub\,(G)$ is the set of all the subgroups of G and R is the set of all right cosets of G, a mapping R$\xrightarrow{\delta}$ $Sub\,(G)$ arises. Thus, $R\delta$ is the unique subgroup of G such that R is a right coset of $R\delta$in G.

In a similar way, if L is the set of all left cosets of G we can define the mapping L$\xrightarrow{\sigma}$ $Sub\,(G)$. If A is a subset of G we define $A^{-1} = \left\{a^{-1}|a \in A\right\}$. It is clear that $\left(A^{-1}\right)^{-1} = A$. In particular, if $R = Hg$ $\in$R, $R^{-1} = \left\{r^{-1}|r = hg, h \in H\right\} = g^{-1}H$. Thus, if R $\in$R, R^{-1} $\in$L.

Let $\mathscr{K}$ be the set of conjugation classes of the subgroups of G. We can define a mapping $Sub\,(G) \xrightarrow{\kappa} \mathscr{K}$ which associates to every subgroup of G its conjugation class. We further introduce the mappings $\mathscr{R} \xrightarrow{\rho} \mathscr{K}$ and $\mathscr{L} \xrightarrow{\lambda} \mathscr{K}$ defined as $\rho = \delta\kappa$ and $\lambda = \sigma\kappa$.

We observe that the position $g \mapsto Rg$ defines an action of G on R, which is equipped in this way with a structure of G-set. Two elements R and R' belong to the same orbit if and only if $R\delta = R'\delta$. If $R = R\delta g$, the stabilizer of R is $g^{-1}R\delta g$.

Let X be a G-set. We can partition it in orbits. Let Ω be the set of its orbits. Each orbit $\omega \in \Omega$ is a homogeneous G-set. The stabilizers of all the elements of a given orbit are mutually conjugate, so that, if Γ *is the set of the conjugation classes of the subgroups of* G, we can assign to each orbit ω a conjugation class $\kappa \in \Gamma$. In this way, a mapping $\Omega \xrightarrow{\varsigma} \Gamma$ arises. We can associate to each $\kappa \in \Gamma$ the cardinality of its inverse image $\kappa\varsigma^{-1}$. Thus, we obtain a surjective mapping ϖ from Γ to the set of cardinals corresponding to some κ. We will call the mapping ϖ the *symmetry spectrum* of X.

We can prove that *two G-sets X and X' are isomorphic if and only if* $\varpi = \varpi'$.

For the necessity, we observe that, if j is an isomorphism, the mapping $\omega \mapsto \omega j$ establishes a bijection between Ω and Ω'. Furthermore, $H_x = H_{xj}$, so that $\omega\varsigma = \omega j\varsigma'$ and hence $\varsigma = j\varsigma'$. We have $\Omega\varsigma = \Omega'\varsigma'$. If $\kappa \notin \Omega\varsigma, \kappa \notin \Omega'\varsigma'$ and vice versa, so that $\kappa\varpi = \kappa\varpi' = 0$, while if $\kappa \in \Omega\varsigma = \Omega'\varsigma'$ $\kappa\varsigma'^{-1} = \kappa\varsigma^{-1}j$, so that $\kappa\varpi = \kappa\varpi'$.

Conversely, suppose that $\varpi = \varpi'$. Let $\Gamma_1 = \{\kappa \in \Gamma | \kappa\varpi \neq 0\}$ and $\kappa \in \Gamma_1$. Then $\kappa\varsigma^{-1} \triangleq \Omega_\kappa$ and $\kappa\varsigma'^{-1} \triangleq \Omega'_\kappa$ are nonempty. Of course the families $\{\Omega_\kappa | \kappa \in \Gamma_1\}$ and $\left\{\Omega'_\kappa | \kappa \in \Gamma_1\right\}$ are partitions of the sets of orbits Ω and Ω', respectively, and they are labeled with injective mappings. As Ω_κ and Ω'_κ have the same cardinality, the set B_κ of bijections from Ω_κ to Ω'_κ is nonempty. Thus we get a family of nonempty sets $\{B_\kappa | \kappa \in \Gamma_1\}$. Defining $B = \underset{\kappa\in\Gamma_1}{\times} B_\kappa$, we select an element $b \in B$, that is, a mapping $b : \kappa \mapsto b_\kappa$ such that $b_\kappa \in B_\kappa$. Each b_κ is a bijection which assigns to each orbit $\omega \in \Omega_\kappa$ the orbit $\omega b_\kappa \in \Omega'_\kappa$. As ω and ωb_κ have the same conjugation class of stabilizers, the set Φ_ω of isomorphisms from ω to ωb_κ is nonempty. If we define $\Phi_\kappa = \underset{\omega\in\Omega_\kappa}{\times} \Phi_\omega$, we select an element $\varphi_\kappa \in \Phi_\kappa$, that is, a mapping $\varphi_\kappa : \omega \mapsto \tilde{\varphi}_\omega \,|\, \omega \in \Omega_\kappa$ such that $\tilde{\varphi}_\omega \in \Phi_\omega$. For each orbit $\omega \in \Omega$ there is a unique $\kappa \in \Gamma_1$ such that $\omega \in \Omega_\kappa$. We define $\varphi_\omega = \tilde{\varphi}_\omega$. Its codomain is $\omega b_\kappa \in \Omega'_\kappa$. For each $x \in X$ there is a unique $\omega \in \Omega$ such that $x \in \Omega$. We can associate to x the element $x\tilde{\varphi}_\omega$. In this way, we get a mapping $X \xrightarrow{\varphi} X'$ which is an injective morphism. As the codomain of φ is the union of the family of the codomains of all the $\tilde{\varphi}_\omega$, it coincides with the union of the family of all orbits each belonging to some set Ω'_κ for all $\kappa \in \Gamma_1$, that is, to all orbits and hence with X'. We conclude that φ is an isomorphism.

Let X be a G-set and Ω its partition in orbits. Putting $K_\omega = \omega\varsigma \in \Gamma$, we consider the set $K \triangleq \underset{\omega\in\Omega}{\times} K_\omega$. We select an element $\psi \in K$, that is, a mapping $\omega \mapsto H_\omega$ such that $H_\omega \in K_\omega$ and define $X' = \bigcup_{\omega\in\Omega} \{\omega\} \times G/H_\omega$. We introduce in X' the action $(\omega, R_\omega)\, g \triangleq (\omega, R_\omega g)$. The elements (ω_2, R_2) and (ω_1, R_1) belong to the same orbit if and only if there is $g \in G$ such that $(\omega_2, R_2) = (\omega_1, R_1)\, g$. Thus any orbit ω' has the form $\omega' = \{\omega\} \times G/H_\omega$. Hence, the above definition of X' is also its orbit decomposition.

We now define $O = \underset{\omega\in\Omega}{\times} \omega$ and select an element $o \in O$, that is, a mapping $\omega \mapsto o_\omega$ such that $o_\omega \in \omega$ and for each ω we introduce the standard isomorphism $\omega \xrightarrow{\gamma_{o_\omega}} G/H_\omega$. For each $x \in X$ there is a unique $\omega \in \Omega$ such that $x \in \omega$.

We define a mapping $X \xrightarrow{\varphi} X'$ through the rule $x\varphi = \left(\omega, x\gamma_{o_\omega}\right)$. The mapping just introduced is clearly an isomorphism between X and X'. We emphasize that this isomorphism is in no way canonical, because X' depends on ψ and, given ψ, φ depends on o. It is easy to check that $\varpi' = \varpi$, as it must be.

Let $S \xrightarrow{\iota} X$ be a free G-set on S and $\sim$ a congruence relation on X. The bijection $(s, g) \mapsto (s\iota)\, g$ allows the transfer of the congruence in the free G-set $S \times G$, so that, without loss of generality, we can put $X = S \times G$ and $\iota : s \mapsto (s, e)$. Let $X \xrightarrow{\pi} X/\!\sim$ be the canonical projection on the quotient. The congruence $\sim$ is the kernel of π as well as the kernel of any morphism of the form πj, being j an arbitrary isomorphism.

Let Ω be the set of orbits of $X/\!\sim$, K_ω the conjugation class of the orbit ω, $K = \underset{\omega\in\Omega}{\times} K_\omega$, and $\psi \in K$. We introduce the G-set $X' = \bigcup_{\omega\in\Omega} \{\omega\} \times G/\bar{H}_\omega$ where $\bar{H}_\omega \triangleq \omega\psi$.

Let φ be an isomorphism between $X/\!\sim$ and X', and put $\chi = \pi\varphi$. $X \xrightarrow{\chi} X'$ is an epimorphism and $\sim = \ker(\chi)$. Hence $x \sim x'$ is and only if $x\chi = x'\chi$.

As X is free, each $x \in X$ is uniquely expressed as $x = (s\iota)\, g$. Defining the mapping $S \xrightarrow{\vartheta} X'$ as $\vartheta = \iota\chi$ we have $x\chi = (s\iota)\, g\chi = s\vartheta g$. X' is a subset of the Cartesian product $\Omega \times \mathrm{R}$. We can define the mappings $X' \xrightarrow{p_1} \Omega$ and $X' \xrightarrow{p_2} \mathrm{R}$ as the restrictions to X' of the projections of the Cartesian product on Ω and R, respectively.

We introduce the mappings $S \xrightarrow{f_1} \Omega$ and $S \xrightarrow{f_2} \mathrm{R}$ defined as $f_1 = \vartheta p_1$ and $f_2 = \vartheta p_2$. We have $s\vartheta = (sf_1, sf_2)$. Hence, $X\chi = \{(sf_1, sf_2 g)\, | s \in S, g \in G\}$. As χ is surjective, $f_1 = \vartheta p$ is surjective. Therefore, we can introduce a partition Σ of S such that $f_1 = qv$, where q is the canonical projection on Σ and v is a bijection $\Sigma \xrightarrow{v} \Omega$. For every $Q \in \Sigma$ we introduce $H_Q \triangleq \bar{H}_{Qv}$.

As $s\vartheta = (sf_1, sf_2) \in X'$, sf_2 belongs to $G/\bar{H}_{sf_1} = G/H_{sq}$. If we put $sq = Q_s$, we conclude that $(s, g) \sim \left(s', g'\right)$ if and only if $Q_s = Q_{s'}$ and $R_s g = R_{s'} g'$, with R_s and $R_{s'}$ right cosets of H_{Q_s}. We define $S \xrightarrow{\phi} \mathrm{R}$ through the rule $s \mapsto R_s$, that is, $\phi = f_2$. As $R_s\delta = H_{Q_s}$, $Q_s = Q_{s'} \Rightarrow R_s\delta = R_{s'}\delta$. Let Σ_ϕ be the partition associated to $\phi\delta$. Then the partition Σ is a *refinement* of $\Sigma_{\phi\delta}$ (written $\Sigma \leq \Sigma_{\phi\delta}$). With these definitions, we can say that $(s, g) \sim \left(s', g'\right)$ if and only if $sq = s'q$ and $s\phi g = s'\phi g'$-

Summarizing the above discussion, we can say that for each congruence on the free G-set X over the set S there are a mapping $S \xrightarrow{\phi} \mathrm{R}$ and a partition Σ of S satisfying the condition $\Sigma \leq \Sigma_{\phi\delta}$ such that $x = (s\iota)\, g \sim \left(s'\iota\right) g' = x'$ if and only if $sq = s'q$ and $s\phi g = s'\phi g'$.

We observe that the condition $\Sigma \leq \Sigma_{\phi\delta}$ is equivalent to the fact that q is a left factor of $\phi\delta$, that is, $\phi\delta = qr$ with $\Sigma \xrightarrow{r} Sub\,(G)$. The mapping r is uniquely determined by $\phi\delta$ and q. Indeed, each $Q \in \Sigma$ can be expressed as $Q = sq$ for some $s \in S$, so that $Qr = sqr = s\phi\delta$. Thus we conclude that

For each congruence on the free G-set X over the set S, there are a mapping $S \xrightarrow{\phi} \mathrm{R}$ *and a projection* q *of* S *on a quotient set* Σ *satisfying the condition that* q *is a left factor of* $\phi\delta$ *such that* $x = (s\iota)\, g \sim \left(s'\iota\right) g' = x'$ *if and only if* $sq = s'q$ *and* $s\phi g = s'\phi g'$.

Conversely, *suppose that a mapping* $S \xrightarrow{\phi} \mathrm{R}$ *and a projection* q *of* S *on a quotient set* Σ *satisfying the condition that* q *is a left factor of* $\phi\delta$ *are given. Then the relation* $x = (s\iota)\, g \sim \left(s'\iota\right) g' = x'$ *if and only if* $sq = s'q$ *and* $s\phi g = s'\phi g'$ *is a congruence.*

Indeed the above relation is an equivalence relation. It is then sufficient to prove that $(s\iota)\, g \sim \left(s'\iota\right) g' \Rightarrow (s\iota)\, g\bar{g} \sim \left(s'\iota\right) g'\bar{g}$ for every $\bar{g} \in G$.

Suppose that $(s\iota)\, g \sim \left(s'\iota\right) g'$. Then $sq = s'q$ and $s\phi g = s'\phi g'$. Given $\bar{g} \in G$, we have $s\phi\,(g\bar{g}) = (s\phi g)\,\bar{g} = \left(s'\phi g'\right)\bar{g} = s'\phi\left(g'\bar{g}\right)$, so that $x\bar{g} = (s\iota)\,(g\bar{g})$ and $x'\bar{g} = \left(s'\iota\right)\left(g'\bar{g}\right)$ with $sq = s'q$ and $s\phi\,(g\bar{g}) = s'\phi\left(g'\bar{g}\right)$, whence $x\bar{g} \sim x'\bar{g}$.

Introducing the mapping $X \xrightarrow{\zeta} \Sigma \times \mathrm{R}$: $(s, g) \mapsto (sq, s\phi g)$, the classes of congruence are the inverse images $(Q, R)\,\zeta^{-1}$ of the elements of the image of ζ. The latter is the set of pairs (Q, R) for which the equations $sq = Q$, $s\phi g = R$ have a solution for (s, g) and, for each such pair (Q, R), the corresponding class of congruence is the set of solutions of these equations.

We have $s\phi g\delta = s\phi\delta = R\delta$, whence $sqr = R\delta$, so that $Qr = R\delta$. Conversely, if this condition is satisfied, putting $Qr = H_Q$ and $s\phi = R_s$, the equations become $sq = Q$ and $R_s g = R$ with $R\delta = H_Q$ and $R_s\delta = H_Q$. The general solution of the first equation is $s \in Q$. Let us express the general solution for g of the equation $R_s g = R$. Suppose that $R_s = H_Q\bar{g}_s$ and $R = H_Q\bar{g}$, so that $H_Q\bar{g}_s g = H_Q\bar{g}$ and then $g \in \bar{g}_s^{-1} H_Q\bar{g} = \bar{g}_s^{-1} H_Q H_Q\bar{g}$. Thus, g is a solution if and only if $g \in R_s^{-1}R$. Hence, the image of ζ is the set P of pairs (Q, R) such that $Qr = R\delta$ and, if $C_{(Q,R)}$ is the congruence class of the pairs (s, g) labeled by $(Q, R) \in P$, we get $C_{(Q,R)} = \left\{(s, g)\,|s \in Q, g \in R_s{}^{-1}R\right\}$.

Putting $R_s^{-1} = L_s$, we can represent $C_{(Q,R)}$ in the form

$$C_{(Q,R)} = \bigcup_{s \in Q} \{s\} \times L_s R \tag{A.2.1}$$

with $Q \in \ker(q)$, $\phi\delta = qr$, $Qr = R\delta$ and $L_s = (s\phi)^{-1}$.

The given congruence can be characterized by a mapping which associates to each (s', g) its class of congruence $C_{(s',g)}$. Then there is a unique pair (Q, R) in the given representation such that $C_{(s',g)} = C_{(Q,R)}$. We have $Q = s'q$ and $g \in L_{s'}R$. If $L_{s'} = g'H_Q$, we have $e \in g^{-1}g'R$ and then $R = H_Q g'^{-1} g = L_{s'}^{-1} g = R_{s'} g$. Thus, we get

$$C_{(s',g)} = \bigcup_{s \in s'q} \{s\} \times R_s^{-1} R_{s'} g, \tag{A.2.2}$$

where $R_s = s\phi$ and with the condition that q is a left factor of $\phi\delta$.

Putting $g = e$ in (A.2.2) and calling p_S be the projection of $S \times G$ on the first factor, we get $[C_{(s',e)} = \bigcup_{s \in Q} \{s\} \times R_s^{-1} R_{s'}$ and then $C_{(s',e)} p_S = \bigcup_{s \in Q} \{s\} = Q$. Hence, the partition Σ is uniquely determined by the congruence and the mapping q with it. Thus, two representations of the same congruence must give rise to the same partition.

We also have $C_{(s',e)} \bigcap (s p_S^{-1}) = \{s\} \times R_s^{-1} R_{s'}$ for every $s \in C_{(s',e)} p_S = Q$. If p_G denotes the projection of $S \times G$ on the second factor, we get $R_s^{-1} R_{s'} = (C_{(s',e)} \bigcap (s p_S^{-1}))\, p_G$ for $s, s' \in Q$ with $R_s \delta = H_Q$. These equations must be satisfied for every representation of the given congruence. Thus, if $\bar{R}_s$ with $\bar{R}_s \delta = \bar{H}_Q$ and R_s with $R_s \delta = H_Q$ correspond to two representations of the same congruence, we must have $R_s^{-1} R_{s'} = \bar{R}_s^{-1} \bar{R}_{s'}$ for $s, s' \in Q$. Let us represent $\bar{R}_s$ as $\bar{R}_s = \bar{H}_Q \bar{g}_s$ and R_s as $R_s = H_Q g_s$. Thus, we have $g_s^{-1} H_Q g_{s'} = \bar{g}_s^{-1} \bar{H}_Q \bar{g}_{s'}$. If we take $s' = s$, we see that H_Q and $\bar{H}_Q$ belong to the same conjugation class, so that there is $\hat{g} \in G$ such that $H_Q = \hat{g} \bar{H}_Q \hat{g}^{-1}$. Thus, we have $g_s^{-1} \hat{g} \bar{H}_Q \hat{g}^{-1} g_s = \bar{g}_s^{-1} \bar{H}_Q \bar{g}_s$, i.e., $\bar{g}_s g_s^{-1} \hat{g} \bar{H}_Q \hat{g}^{-1} g_s \bar{g}_s^{-1} = \bar{H}_Q$. Therefore $\hat{g}^{-1} g_s \bar{g}_s^{-1} \in \bar{N}_Q$ where $\bar{N}_Q$ is the normalizer of $\bar{H}_Q$. Hence we get $g_s = \hat{g} n_s \bar{g}_s$ with $n_s \in \bar{N}_Q$. Then we obtain $R_s = \hat{g} \bar{H}_Q \hat{g}^{-1} \hat{g} n_s \bar{g}_s = \hat{g} \bar{H}_Q n_s \bar{g}_s = \hat{g} n_s \bar{H}_Q \bar{g}_s = \hat{g} n_s \bar{R}_s$. We further have $R_s^{-1} R_{s'} = \bar{R}_s^{-1} n_s^{-1} n_{s'} \bar{R}_{s'} = \bar{R}_s^{-1} \bar{R}_{s'}$. Consequently $\bar{g}_s^{-1} \bar{H}_Q n_s^{-1} n_{s'} \bar{H}_Q \bar{g}_{s'} = \bar{g}_s^{-1} \bar{H}_Q \bar{g}_{s'}$, i.e., $\bar{H}_Q n_s^{-1} n_{s'} = \bar{H}_Q$. This means that $n_s \bar{H}_Q = n_{s'} \bar{H}_Q$, that is, that the element $n_s \bar{H}_Q$ of $\bar{N}_Q / \bar{H}_Q$ is independent of s for $s \in Q$. Thus, there is $n \in \bar{N}_Q$ such that $n_s \bar{H}_Q = n \bar{H}_Q$. Consequently, $n_s = n h_s$ with $h_s \in \bar{H}_Q$. Thus $R_s = \hat{g} n_s \bar{R}_s = \hat{g} n \bar{R}_s$. Putting $g_Q = \hat{g} n$, we have $R_s = g_Q \bar{R}_s$, which entails $H_Q = g_Q \bar{H}_Q g_Q^{-1}$. Conversely, if $R_s = g_Q \bar{R}_s$, $R_s^{-1} R_{s'} = \bar{R}_s^{-1} g_Q^{-1} g_Q \bar{R}_{s'} = \bar{R}_s^{-1} \bar{R}_{s'}$. As q and ϕ uniquely determine the congruence, we conclude the following.

The mappings ϕ, q and ϕ', q' such that q is a left factor of $\phi\delta$ and q' is a left factor of $\phi'\delta$ define the same congruence if and only if $q' = q$ and there is a mapping $S/\ker(q) \xrightarrow{\mu} G$ factorizable by q on the left such that $s\phi' = (sq\mu)(s\phi)$.

Returning to the initial free G-set $S \xrightarrow{\iota} X$ we can rewrite (A.2.2) in the form

$$C_{(s'\iota)g} = \bigcup_{s \in s'q} (s\iota)(s\phi)^{-1} (s'\phi)\, g. \tag{A.2.3}$$

The mapping q is completely determined by the congruence, while ϕ is determined by the congruence up to an arbitrary pointwise left multiplier $q\mu$ with values in G.

Consider the special case when Σ consists of singletons. In this case, (A.2.3) becomes $C_{(s\iota)g} = (s\iota)(s\phi)^{-1}(s\phi)\, g$. We can express $s\phi$ as $s\phi = (s\phi\delta)\, g_s$, so that $(s\phi)^{-1}(s\phi) = g_s^{-1}(s\phi\delta)\, g_s$ which is a group H_s independent of the choice of ϕ for the given congruence. Thus, in this special case, the congruence is uniquely represented through a mapping $s \mapsto H_s$ with values in $Sub\,(G)$. We have $C_{(s\iota)g} = (s\iota)\, H_s g$.

A.2.3 Generators. Frame

We say that a subset T of a G-set Y is a *free set of generators* for Y if $T \stackrel{in}{\to} Y$ is a free G-set. We say that Y is freely generated if it admits a free set of generators. A G-set is freely generated if and only if it is isomorphic to the codomain of a free G-set. Indeed, if Y is freely generated, there is a free set T of generators, so that $T \stackrel{in}{\to} Y$ is a free G-set and $Y \stackrel{id}{\to} Y$ is an isomorphism. Conversely, suppose that Y is isomorphic to the codomain of a free G-set $S \stackrel{\iota}{\to} X$. Let $X \stackrel{\eta}{\to} Y$ be an isomorphism. Put $T = S\iota\eta$. Then $T \stackrel{in}{\to} Y$ is a free G-set. Indeed, let $T \stackrel{f}{\to} Z$ be a mapping in a G-set. Let $\beta\, in$ be the canonical decomposition of $\iota\eta$ and define $f' = \beta f$. There is a unique morphism $X \stackrel{\varphi'}{\to} Z$ such that $f' = \iota\varphi'$. Define $\varphi = \eta^{-1}\varphi'$. We have $in\eta^{-1} = \beta^{-1}\iota$, so that $in\varphi = \beta^{-1}\iota\varphi' = \beta^{-1}f' = f$. The morphism φ is unique. Indeed, if $in\bar{\varphi} = f$, $\bar{\varphi}' \triangleq \eta\bar{\varphi}$ satisfies $f' = \iota\bar{\varphi}'$ so that $\bar{\varphi}' = \varphi'$ and hence $\bar{\varphi} = \varphi$.

A free set of generators of the freely generated G-set Y will also be called a *frame*.

Suppose that Y is freely generated and that T is a free set of generators of Y. As $T \stackrel{in}{\to} Y$ is a free G-set, any $y \in Y$ is expressed in a unique way as $y = tg$. Thus, there is a bijection $T \times G \stackrel{F_T}{\to} Y : (t, g) \mapsto tg$. We call (t, g) the T-coordinates of y with respect to the frame T. $T \stackrel{\iota}{\to} T \times G$ and $T \stackrel{in}{\to} Y$ are free, so that there is a unique isomorphism from $T \times G$ to Y that extends the mapping defined by $(t, e) \mapsto in \cdot t = t$. As $(t, e)\, F_T = t$, F_T is this unique morphism. This morphism is called the T-coordinate morphism of Y with respect to the frame T. Conversely, suppose that $T \subseteq Y$ is such that each $y \in Y$ is expressed in a unique way as $y = tg$ with $(t, g) \in T \times G$. The isomorphism $(t, g) \mapsto tg$ extends the mapping $(t, e) \mapsto in \cdot t = t$, so that $T \stackrel{in}{\to} Y$ is a free G-set. Thus we conclude the following.

A G-set Y is freely generated if and only if there is $T \subseteq Y$ such that any $y \in Y$ is expressed in a unique way as $y = tg$ with $(t, g) \in T \times G$.

Let Ω be the set of orbits of Y. If $\omega \in \Omega$, ωF_T^{-1} is an orbit of $T \times G$ and conversely. But we know that all the orbits of $T \times G$ are torsors. As Y is an isomorphic image of $T \times G$, each orbit of Y is a torsor. Conversely, suppose that each orbit of a G-set is a torsor. Consider the canonical projection $Y \stackrel{\pi}{\to} \Omega$. Let $\Omega \stackrel{\sigma}{\to} Y$ be a section of π. We prove that $T \triangleq \Omega\sigma$ is a free set of generators for Y Indeed, let $y \in Y$. The element y and $y\pi\sigma$ belong to the same orbit. Hence there is $g \in G$ such that $y = (y\pi\sigma)\, g$. Putting $t = y\pi\sigma \in T$ we can write $y = tg$. Let $y = t'g'$ with $t' \in T$. Thus $t' = \omega'\sigma$

for some $\omega' \in \Omega$ and then $y = \omega'\sigma g'$. But $y\pi = \omega'$, so that $t = y\pi\sigma = \omega'\sigma = t'$. Thus $y = tg = tg'$. As every orbit is a torsor, $g' = g$. Thus y is uniquely expressed as tg with $(t, g) \in T \times G$, so that Y is a freely generated G-set. Thus we conclude the following.

The G-set Y is freely generated if and only if its every orbit is a torsor.

Furthermore, the above argument proves the following.

If the G-set Y is freely generated and $Y \xrightarrow{\pi} \Omega$ is the canonical projection of Y on the set Ω of its orbits, the image of every section of π is a set of free generators for Y.

We now prove the following.

If the G-set Y is freely generated and $Y \xrightarrow{\pi} \Omega$ is the canonical projection of Y on the set Ω of its orbits, $T \subseteq Y$ is a frame if and only if it is the image of a section of π. Two sections define the same frame if and only if they are equal. The restriction τ of π to T is bijective.

The sufficiency has already been proved. Conversely, suppose that T is a free set of generators. Let $\omega \in \Omega$. Then there is y such that $\omega = y\pi$. But y admits a unique representation $y = tg$. Let $\omega = y'\pi$ and $y' = t'g'$. Furthermore, $y' = y\bar{g}$ for some $\bar{g} \in G$. Thus $y = t'g'\bar{g}^{-1}$ and, for the uniqueness of the representation, $t' = t$. In this way, a well-defined mapping $\Omega \xrightarrow{\sigma} Y$ arises which associates t to ω. We have $\omega = y\pi = tg\pi = t\pi = \omega\sigma\pi$. Thus $\sigma\pi = id$ and σ is a section of π.

Furthermore $\Omega\sigma \subseteq T$. Let $\bar{t} \in T$ and $\omega = \bar{t}\pi$. Let us evaluate $\omega\sigma$. As $\omega = \bar{t}\pi$, we can take $y = \bar{t}$ in the equation $y = tg$ which defines σ. Thus, we have $\bar{t} = (\omega\sigma)\, g$ which has $(\bar{t}, e)$ as its unique solution. In particular, $\bar{t} = \omega\sigma$, so that $T = \Omega\sigma$. Thus the necessity is proved.

Let $\Omega\sigma = \Omega\sigma'$. Suppose that $\omega \in \Omega$. Then $t = \omega\sigma \in \Omega\sigma'$. Thus $t = \omega'\sigma'$, $\omega' \in \Omega$. We have $t\pi = \omega$, $t\pi = \omega'$, so that $\omega\sigma = \omega\sigma'$ and $\sigma = \sigma'$. Furthermore $\omega\sigma\pi = \omega$ and $t\pi\sigma = t$. If υ is the restriction of σ to its image, this means that τ and υ are inverse to each other.

From the above discussion, we can conclude that the following sentences are equivalent:

1) *Y is a freely generated G-set,*

2) *every orbit of Y is a torsor, and*

3) *Y is a disjoint union of torsors.*

The equivalence of 1) and 2) has already been proved. Furthermore 2)⇒3) because the orbits are mutually disjoint. Let Y be a disjoint union of torsors. If $y \in Y$, it belongs to a unique torsor of the union. This torsor is a homogeneous G-set and hence is an orbit. Therefore, every orbit of Y is a torsor.

If T' is a new frame of Y, each given element $y \in Y$ has T-coordinates (t', g') in the new frame, such that $y = tg = t'g'$. We obtain a new T-coordinate morphism $T' \times G \xrightarrow{F_{T'}} Y$, so that a bijection $T \times G \xrightarrow{F_T F_{T'}^{-1}} T' \times G$ arises. The coordinate transformation which expresses the new coordinates of y in terms of the old ones is

$(t, g)\, F_T F_{T'}^{-1} = \left(t', g'\right)$. Let τ, τ' be the restrictions of $Y \xrightarrow{\pi} \Omega$ to the frames T, T', respectively. We can write $y\pi = t\pi = t\tau = t'\pi = t'\tau'$. Thus $t' = t\tau\tau'^{-1}$. Putting $f_1 = \tau\tau'^{-1}$, we can write $t' = t f_1$. We observe that $T \xrightarrow{f_1} T'$ is a bijection. Given t, the equation $tg = (t f_1)\, g'$ defines a mapping $g \mapsto g'$. As $tg\bar{g} = (t f_1)\, g'\bar{g}$, for the uniqueness of the representation $g\bar{g} \mapsto g'\bar{g}$. Thus, introducing the mapping $T \xrightarrow{f_2} G$ uniquely defined by the position $t = (t f_1)\,(t f_2)$, we can write $g' = (t f_2)\, g$. Hence $(t, g)\, F_T F_{T'}^{-1} = (t f_1, (t f_2)\, g)$. We conclude as given below.

If T, T' are frames of a freely generated G-set Y, the T-coordinates of any element of Y in the frame T' are expressed in terms of the coordinates of the same element in the frame T by means of the transformation $t' = t f_1$ and $g' = (t f_2)\, g$, where $f_1 = \tau\tau'^{-1}$ and f_2 is uniquely defined by the position $t = (t f_1)\,(t f_2)$.

There is a natural bijection between the frame T and the set Ω. This bijection is $T \xrightarrow{\tau} \Omega$. Thus, we can use (ω, g) as the coordinates of an element of Y rather than (t, g). Such coordinates will be called orbit coordinates (shortly Ω-coordinates). The element y is expressed in terms of these coordinates as $y = tg = (\omega\sigma)\, g$. The transformation from (ω, g) to (t, g) is given by $(\omega, g) \mapsto (\omega\upsilon, g)$, and the inverse transformation is $(t, g) \mapsto (t\tau, g)$. The transformation rule for orbit coordinates from the frame T to the frame T' can be obtained by transformation composition. We have

$$(\omega, g) \mapsto (\omega\upsilon, g) \mapsto (\omega\upsilon f_1, (\omega\upsilon f_2)\, g) \mapsto \left(\omega\upsilon f_1\tau', (\omega\upsilon f_2)\, g\right).$$

But $\upsilon f_1\tau' = \upsilon\tau\tau'^{-1}\tau' = id$. Thus, putting $\upsilon f_2 = f$, we have the transformation law $\omega' = \omega$, $g' = (\omega f)\, g$, with $\Omega \xrightarrow{f} G$.

The frame T is represented in T-coordinates by the set $T \times \{e\}$, and in Ω-coordinates by $\Omega \times \{e\}$. Hence, the frame T is represented in the frame T' in Ω-coordinates by $\{(\omega, \omega f) \mid \omega \in \Omega\}$. Thus we conclude the following.

If T, T' are frames of Y and (ω, g), $\left(\omega', g'\right)$ are the Ω-coordinates of a point of Y in T and T', respectively, there is a unique mapping $\Omega \xrightarrow{f} G$ such that $\omega' = \omega$, $g' = (\omega f)\, g$. This mapping is defined by the condition that ωf represents the shift from the point over ω of frame T' to the point over ω of frame T. Consequently, the corresponding sections are transformed according to the rule $\sigma = \sigma' f$, where the action is defined pointwise.

We note that, while the domain of T-coordinates is frame-dependent, the domain of Ω-coordinates is not, and it is given by $\Omega \times G$.

Henceforth, the Ω-coordinates will be used in most cases, so that we will call them simply coordinates.

Appendix B
Barycentric Coordinates

B.1 Simplexes and Systems of Linear Inequalities

Let us consider[1] a finite set E of cardinality n. We suppose that a total order is specified on it, so that we can attach to each element of E a label belonging to $\mathbb{N}_n \triangleq \{i \in \mathbb{N} \,|\, i \leq n\}$.

We interpret E as the set of vertices of an $(n-1)$-*dimensional simplex* T_{n-1}. Introducing the barycentric coordinates, a point P of the simplex will be represented as $P = \sum\limits_{i=1}^{n} p_i e_i$ where $p_i \geq 0$ and $\sum\limits_{i=1}^{n} p_i = 1$.

Consider the subset S_{n-1} of T_{n-1} defined by the inequalities $p_1 \geq p_2 \geq .. \geq p_n$, which is a convex set. Its extreme points are obtained by taking $n-1$ among the $2n-1$ inequalities $p_1 \geq p_2 \geq .. \geq p_n$ and $p_i \geq 0$ with equality sign in all possible way consistent with them.

If we take all the equalities in the first group, we get $p_i = 1/n$. If we take at least an equality in the second group, there will be a smallest index $k+1$ for which $p_{k+1} = 0$.

Of course, observing that at least one of the p_i must be nonzero, we must have $k \geq 1$. This entails $p_{k+2} = .. = p_n = 0$. Hence, there are $n-k+1$ equalities of this kind. The remaining $n-1-(n-k) = k-1$ must belong to the first group. Thus, we must have $p_1 = p_2 = .. = p_k$ and $p_{k+1} = .. = p_n = 0$. Hence we have $p_i = 1/k$ for $i \leq k$ and $p_i = 0$ for $i > k$.

Thus, the extreme points of S_{n-1} are $P_k = \sum\limits_{i=1}^{k} \frac{1}{k} e_i$. Their number is n and they are independent. Therefore, is simple, so that

$$S_{n-1} = \left\{ \sum_{k=1}^{n} \lambda_k P_k \mid \lambda_k \geq 0, \sum_{k=1}^{n} \lambda_k = 1 \right\}.$$

[1]The content of this chapter can be found in the Campanella's file bary (29-03-2014)

M. Campanella et al., *Interpretative Aspects of Quantum Mechanics*, UNIPA Springer Series,
https://doi.org/10.1007/978-3-030-44207-1

Let B denote the set of forms of the type $\sum_{i=1}^{n} a_i p_i$ with $a_i \in \{0, 1\}$.

If b, b' belong to B, consider the inequality $b \leq b'$. The set of solutions of this inequality belonging to S_{n-1} may be empty, or it may be a proper subset of S_{n-1}, or else it may be the whole S_{n-1}. In the last case, we say that $b \leq b'$ is a *theorem*. In the first case, we say that $b \leq b'$ is a *contradiction*, while in the second case we say that $b \leq b'$ is an *axiom*.

Let $b = \sum_{i=1}^{n} a_i p_i$ and $b' = \sum_{i=1}^{n} a'_i p_i$. The inequality $b \leq b'$ is equivalent to $\sum_{i=1}^{n} c_i p_i \geq 0$ with $c_i = a'_i - a_i \in \{-1, 0, 1\}$. Conversely, an inequality of the latter form is always equivalent to an inequality of the form $b \leq b'$. Indeed, for each $i \in \mathbb{N}_n$ put $a_i = 0$ if $c_i \geq 0$ and $a_i = 1$ otherwise. Similarly, we put $a'_i = 0$ if $c_i \leq 0$ and $a'_i = 1$ otherwise.

Thus, for each class of equivalence of inequalities, there is a unique representative in which no term on the left side appears in the right side. This unique representative will be called the *reduced form* of the inequality. *An inequality is a theorem (a contradiction, an axiom) if and only if its reduced form is a theorem (a contradiction, an axiom).* We observe that the reduced forms are in bijection with the sequences c_i.

The restriction to S_{n-1} of the form $\sum_{i=1}^{n} c_i p_i$ can be expressed using the barycentric coordinates of S_{n-1}. We get $\sum_{i=1}^{n} c_i p_i = \sum_{k=1}^{n} \frac{\lambda_k}{k} z_k$ with $z_k = \sum_{i=1}^{k} c_i$.

If the inequality is a contradiction, this expression must take negative values in all the points of S_{n-1}. In particular, it must take negative values on the vertices, so that we must have $z_k < 0$ for every $k \in \mathbb{N}_n$. This condition is clearly also sufficient because the λ_k are nonnegative and at least one is nonzero. We say that the sequence z_k is *strictly definite negative*.

Suppose now that the inequality is a theorem. In this case the form must take nonnegative values on the vertices. Thus we must have $z_k \geq 0$ for every $k \in \mathbb{N}_n$. This condition is clearly also sufficient because the λ_k are nonnegative. We say that the sequence z_k is *semidefinite positive*.

Clearly the inequality is an axiom if and only if it is neither a contradiction nor a theorem. Thus there must be $h \in \mathbb{N}_n$ such that $z_h < 0$ and $k \in \mathbb{N}_n$ such that $z_k \geq 0$. Thus, the sequence is indefinite or semidefinite negative, but not strictly.

Thus

$b \leq b'$ theorem $\Leftrightarrow$ z semidefinite positive

$b \leq b'$ contradiction $\Leftrightarrow$ z strictly definite negative

$b \leq b'$ axiom $\Leftrightarrow$ z indefinite or semidefinite negative, but not strictly.

If we exchange the roles of b and b', z is changed in its opposite, so that we have

z indefinite $\Leftrightarrow$ both $b \leq b'$ and $b' \leq b$ axioms

z semidefinite positive but not strictly $\Leftrightarrow$ $b \leq b'$ theorem and $b' \leq b$ axiom

z strictly definite positive $\Leftrightarrow$ $b \leq b'$ theorem and $b' \leq b$ contradiction

z strictly definite negative $\Leftrightarrow$ $b \leq b'$ contradiction and $b' \leq b$ theorem

z semidefinite negative but not strictly $\Leftrightarrow$ $b \leq b'$ axiom and $b' \leq b$ theorem

In what follows, we consider strict inequalities. This means that from the set of solutions of an inequality we exclude the boundary. Thus, we will exclude from our considerations the case of semidefinite forms.

We have:

z indefinite $\Leftrightarrow$ both $b \leq b'$ and $b' \leq b$ axioms

z strictly definite positive $\Leftrightarrow$ $b \leq b'$ theorem and $b' \leq b$ contradiction

z strictly definite negative $\Leftrightarrow$ $b \leq b'$ contradiction and $b' \leq b$ theorem

B.1.1 Consistent Sets in a Simplex

We denote with Λ_{n-1} the standard simplex of $\mathbb{R}^n$. For each $\lambda \in \Lambda_{n-1}$, we have the natural total order in the set $\{b(\lambda) \mid b \in B\}$.

We put

$$b \leq_\lambda b' \qquad \text{if and only if} \qquad b(\lambda) \leq b'(\lambda).$$

This relation is clearly reflexive and transitive, that is, a preorder relation. Furthermore every pair b, b' can be compared, so that it is a total preorder relation. It is a total order relation if and only if λ is not a solution of any equation of the form $b(\lambda) = b'(\lambda)$ for $b \neq b'$.

We can define in Λ_{n-1} an equivalence relation through the position

$$\lambda \sim \lambda' \qquad \text{if and only if} \qquad \leq_\lambda = \leq_{\lambda'}. \quad (\lambda, \lambda' \in \Lambda_{n-1})$$

Let C be a class of equivalence of $\sim$. Then C is a convex set.

Indeed, let $\lambda, \lambda' \in C$. Thus $\leq_\lambda = \leq_{\lambda'}$, that is, $b(\lambda) \leq b'(\lambda) \Leftrightarrow b(\lambda') \leq b'(\lambda')$. Let $\lambda'' = \mu\lambda + \nu\lambda'$ with $\mu \geq 0, \nu \geq 0$ and $\mu + \nu = 1$. Then $\lambda'' \in \Lambda_{n-1}$ and $b(\lambda'') = \mu b(\lambda) + \nu b(\lambda')$ and similarly for b'. Let $b(\lambda) \leq b'(\lambda)$. Then $b(\lambda') \leq b'(\lambda')$ and $b(\lambda'') = \mu b(\lambda) + \nu b(\lambda') \leq \mu b'(\lambda) + \nu b'(\lambda') = b'(\lambda'')$.

Conversely, suppose that $b(\lambda'') \leq b'(\lambda'')$. Then $\mu(b(\lambda) - b'(\lambda)) + \nu(b(\lambda') - b'(\lambda')) \leq 0$. If $\neg(b(\lambda) \leq b'(\lambda))$, then $\neg(b(\lambda') \leq b'(\lambda'))$, so that both $b(\lambda) - b'(\lambda)$ and $b(\lambda') - b'(\lambda')$ are positive and we get a contradiction.

Let C be a class of equivalence defining a total order. If $\bar{\lambda} \in C$ the relation $\leq_{\bar{\lambda}}$, regarded as a subset of $B \times B$, is

$$S_{\bar{\lambda}} = \left\{ \left(b, b'\right) \in B \times B \mid b\left(\bar{\lambda}\right) \leq b'\left(\bar{\lambda}\right) \right\}.$$

Every pair of different elements can be compared and, being $\leq_{\bar{\lambda}}$ a total order, for every such pair we must have $b\left(\bar{\lambda}\right) < b'\left(\bar{\lambda}\right)$. Let λ be a solution of the set of inequalities

$$\left\{ b(x) < b'(x) \mid \left(b, b'\right) \in S_{\bar{\lambda}},\ b \neq b' \right\}.$$

Then $\lambda \in C$. Indeed, let $b \leq_{\bar{\lambda}} b'$, $b \neq b'$. Then $\left(b, b'\right) \in S_{\bar{\lambda}}$, $b \neq b'$. Consequently, $b(\lambda) < b'(\lambda)$ and $b \leq_{\lambda} b'$. If $b \leq_{\bar{\lambda}} b'$ and $b = b'$, then $b \leq_{\lambda} b'$. Thus $\leq_{\bar{\lambda}} \subseteq \leq_{\lambda}$. As $\leq_{\bar{\lambda}}$ and $\leq_{\lambda}$ are total orders, $\leq_{\bar{\lambda}} = \leq_{\lambda}$ and $\lambda \in C$. Conversely, if $\lambda \in C$, λ is a solution of the set of inequalities $\left\{ b(x) < b'(x) \mid \left(b, b'\right) \in S_{\bar{\lambda}},\ b \neq b' \right\}$.

To each class of equivalence C representing a total order we can associate a set of inequalities S_C defined as follows.

We choose $\bar{\lambda} \in C$ and stipulate $\left(b(x) < b'(x)\right) \in S_C$ if and only if $b\left(\bar{\lambda}\right) < b'\left(\bar{\lambda}\right)$. The definition is well posed, because it is independent of $\bar{\lambda}$. Indeed, let $\lambda \in C$. Thus, we get for this choice the definition $\left(b(x) < b'(x)\right) \in S'_C$ if and only if $b(\lambda) < b'(\lambda)$.

Let $\left(b(x) < b'(x)\right) \in S_C$. Then $b\left(\bar{\lambda}\right) < b'\left(\bar{\lambda}\right)$, so that $b \leq_{\bar{\lambda}} b'$ and $b \neq b'$. As $\lambda \in C$, $b \leq_{\lambda} b'$ and $b(\lambda) \leq b'(\lambda)$. As C defines a total order, we have $b(\lambda) < b'(\lambda)$, whence $\left(b(x) < b'(x)\right) \in S'_C$.

Conversely, suppose that $\left(b(x) < b'(x)\right) \in S'_C$. Then $b(\lambda) < b'(\lambda)$, so that $b \leq_{\lambda} b'$ and $b \neq b'$. As $\bar{\lambda} \in C$, $b \leq_{\bar{\lambda}} b'$ and $b\left(\bar{\lambda}\right) \leq b'\left(\bar{\lambda}\right)$. As C defines a total order, we have $b\left(\bar{\lambda}\right) < b'\left(\bar{\lambda}\right)$, whence $\left(b(x) < b'(x)\right) \in S_C$. It is clear that $\lambda \in C$ if and only if it is a solution of all the inequalities $\left(b(x) < b'(x)\right) \in S_C$.

The set S_C is nonempty, because for every $\lambda \in \Lambda_{n-1}$ $0(\lambda) = 0$ and $1(\lambda) = 1$. Furthermore, as C is nonempty, the set of solutions of the set S_C of inequalities is nonempty.

Suppose now that $\left(b(x) < b'(x)\right) \notin S_C$. Let $\bar{\lambda} \in \Lambda_{n-1}$ be a solution of the set of inequalities $S_C \cup \left\{\left(b(x) < b'(x)\right)\right\}$. As $\bar{\lambda}$ satisfies all the inequalities of S_C, $\bar{\lambda} \in C$. But in addition $b\left(\bar{\lambda}\right) < b'\left(\bar{\lambda}\right)$, so that $\left(b(x) < b'(x)\right) \in S_C$. From the contradiction, we deduce that S_C cannot be included in a larger set of inequalities of the form $b(x) < b'(x)$ with $b \neq b'$ having solutions in Λ_{n-1}.

Let Υ be the set of inequalities defined by

$$\Upsilon = \left\{ b(x) < b'(x) \mid \left(b, b'\right) \in B \times B, b \neq b' \right\}.$$

The set $S_C \subseteq \Upsilon$ has the following properties:

1) $S_C \neq \varnothing$,
2) S_C has solutions in Λ_{n-1}, and
3) if $S' \subseteq \Upsilon$ satisfies 1) and 2) and $S_C \subseteq S'$, then $S' = S_C$.

Conversely, suppose that $S \subseteq \Upsilon$ has the properties 1) –3). Then there is a total order class of equivalence C such that $S_C = S$. Indeed, choosing a solution $\bar{\lambda}$ of the set S, the relation $\leq_{\bar{\lambda}}$ defined by the conditions $b \leq_{\bar{\lambda}} b$ and $b \leq_{\bar{\lambda}} b'$ if and only if $b(\bar{\lambda}) < b'(\bar{\lambda})$ for $b \neq b'$ is a total order. Let C be the class of equivalence of $\leq_{\bar{\lambda}}$. Then the inequality $b(x) < b'(x)$ belongs to S_C if and only if $b(\bar{\lambda}) < b'(\bar{\lambda})$. Consequently, if $(b(x) < b'(x)) \in S$, it belongs to S_C, so that $S \subseteq S_C$. As S is maximal, $S_C = S$.

A subset S of Υ satisfying 1)–3) will be called a *maximal consistent subset.*

We have shown that, if S is a maximal consistent subset, the equation $S_C = S$ has a solution for C in the set of total order classes of equivalence. We now show that there is a unique solution.

Indeed, suppose that $S_C = S_{C'}$ with $C \neq C'$. Let $\bar{\lambda} \in C$. Then $\bar{\lambda}$ is a solution of the set S_C and hence of the set $S_{C'}$. Thus $\bar{\lambda} \in C'$. But this is a contradiction, because different classes of equivalence are disjoint.

We conclude that *there is a bijection between the set of classes of equivalence of total order and the set of maximal consistent subsets of* Υ.

A subset of Υ having solutions in Λ_{n-1} is called a *consistent set.* The empty set is trivially consistent. By definition, a consistent set has a nonempty set of solutions in Λ_{n-1}.

Two consistent sets K and K' are equivalent if they have the same set of solutions. Let K be a consistent set. The class of equivalence of K is a finite set and is composed of finite sets. Hence, in its class of equivalence, there are representatives of minimum cardinality.

A consistent set is called minimal if it is of minimum cardinality in its class of equivalence. A *minimal consistent set* is called a total order minimal consistent set if it is minimal and its class of equivalence contains a maximal consistent subset.

Given a nonempty subset S of Υ the problem to establish whether it is consistent arises.

If the set S is

$$S = \{\alpha_i(x) < \beta_i(x) \mid i \in [1, m]\},$$

it is consistent if the system $\alpha_i(x) < \beta_i(x)$ together with $\sum_{j=1}^{n} x_j = 1$ has nonnegative solutions.

If $\alpha_i(x) = \sum_{j=1}^{n} a_{ij}x_j$ and $\beta_i(x) = \sum_{j=1}^{n} b_{ij}x_j$, putting $c_{ij} = a_{ij} - b_{ij}$, S is consistent if and only if the set of conditions $\sum_{j=1}^{n} c_{ij}x_j < 0, \sum_{j=1}^{n} x_j = 1$ has nonnegative solutions. This happens if and only if the set of strict inequalities $\sum_{j=1}^{n} c_{ij}x_j < 0$ and $\sum_{j=1}^{n} (-x_j) < 0$ has nonnegative solutions. Indeed, let $\xi_j \geq 0\,(j \in [1, n])$ satisfy the conditions $\sum_{j=1}^{n} c_{ij}\xi_j < 0$ and $\sum_{j=1}^{n} \xi_j = 1$. Then $\sum_{j=1}^{n} (-\xi_j) < 0$, so that ξ_j $(j \in [1, n])$ is a nonnegative solution of $\sum_{j=1}^{n} c_{ij}x_j < 0$ and $\sum_{j=1}^{n} (-x_j) < 0$.

Conversely, let $\sum_{j=1}^{n} c_{ij}\xi_j < 0$ and $-a \triangleq \sum_{j=1}^{n} (-\xi_j) < 0$. Then, putting $\xi'_j = \xi_j/a \geq 0$, we get $\sum_{j=1}^{n} c_{ij}\xi'_j < 0$ and $\sum_{j=1}^{n} \xi'_j = 1$, so that the set of conditions $\sum_{j=1}^{n} c_{ij}x_j < 0$, $\sum_{j=1}^{n} x_j = 1$ has nonnegative solutions. Adding the subscript 0 and defining $d_{ij} = c_{ij}$ for $i \neq 0$ and $d_{oj} = -1$, *we conclude that* S *is consistent if and only if the set of inequalities* $\sum_{j=1}^{n} d_{ij}x_j < 0$ *has nonnegative solutions.*

If S is consistent, there are $\xi_j \geq 0\,(j \in [1, m])$ such that $\beta_i \triangleq \sum_{j=1}^{n} d_{ij}\xi_j < 0$. Putting $\beta \triangleq \max\{\beta_i | i \in [0, m]\}$, we have $\sum_{j=1}^{n} d_{ij}\xi_j \leq \beta$. Introducing the quantities $\xi'_i \triangleq \xi_i/|\beta| \geq 0$ we get $\sum_{j=1}^{n} d_{ij}\xi'_j \leq -1$. Hence, the set S' of inequalities $\sum_{j=1}^{n} d_{ij}x_j \leq -1$ has nonnegative solutions. Conversely, suppose that S' has nonnegative solutions. Hence, the set $\sum_{j=1}^{n} d_{ij}x_j < 0$ has nonnegative solutions.

We conclude that S *is consistent if and only if the set of inequalities* $\sum_{j=1}^{n} d_{ij}x_j \leq -1$ *has nonnegative solutions.*

In turn, this set of inequalities is feasible if and only if the maximum of $\zeta = -x_0$ subject to the constraints $\sum_{j=1}^{n} d_{ij}x_j - x_0 \leq -1$, $x_j \geq 0$ for $j \in [0, n]$ is zero. Indeed, suppose that the set of inequalities $\sum_{j=1}^{n} d_{ij}x_j \leq -1$ is feasible. Then there are $\xi_j \geq 0$ $(j \neq 0)$ such that $\sum_{j=1}^{n} d_{ij}\xi_j \leq -1$. The point $(0, \xi_1, ..., \xi_n)$ satisfies all the constraints and ζ attains the value 0 in this point. But, owing to the constraint $x_0 \geq 0$, the maximum attainable value of ζ is not positive. Hence the maximum of ζ is 0. Conversely, suppose that the maximum of $\zeta = -x_0$ subject to the constraints $\sum_{j=1}^{n} d_{ij}x_j - x_0 \leq -1$, $x_j \geq 0$ for $j \in [0, n]$ is zero. This means that there are $\xi_j \geq 0$ $(j \neq 0)$ such that the point $(0, \xi_1, ..., \xi_n)$ satisfies all the constraints, and hence $\sum_{j=1}^{n} d_{ij}\xi_j \leq 0$ and the set of inequalities $\sum_{j=1}^{n} d_{ij}x_j \leq 0$ is feasible.

We conclude that

S is consistent if and only if the maximum of $\zeta = -x_0$ *subject to the constraints* $\sum_{j=1}^{n} d_{ij}x_j - x_0 \leq -1$, $x_j \geq 0$ *for* $j \in [0, n]$ *and* $i \in [0, m]$ *is zero.*

Appendix C
Linear Spaces

C.1 Duals and Adjoints

Let X be a[1] complex linear space. Its dual X^d is by definition the set of all the linear mappings in $\mathbb{C}$, with its natural structure of linear space. If X is finite-dimensional, X^d is isomorphic to X (but no canonical isomorphism is defined). However, there is a canonical isomorphism between X and its double dual X^{dd} which allows the identification of the latter with X. This isomorphism is defined by the position

$$x^{dd}\left(x^d\right) = x^d\left(x\right).$$

If $X \xrightarrow{\varphi} Y$ is a linear mapping, we can define the *transpose mapping*

$$Y^d \xrightarrow{\varphi^d} X^d \qquad \text{as} \qquad \left(\varphi^d y^d\right)(x) = y^d\left(\varphi x\right).$$

We can write $\varphi^d y^d = y^d \varphi$. The rule of calculation

$$(\varphi \circ \psi)^d = \psi^d \circ \varphi^d$$

holds. Furthermore, identifying X with X^{dd}, we have $\left(\varphi^d\right)^d = \varphi$.

If X is a Hilbert space H, the dual of H will be denoted H^*. In this case, a canonical anti-isomorphism * between H and H^* is induced by the scalar product. The canonical anti-isomorphism will be called *conjugation*. In the bra-ket notation, the conjugate of $|x\rangle$ will be denoted $\langle x|$ and will be called the bra conjugate to $|x\rangle$. Hence we write $|x\rangle^* = \langle x|$.

[1] The content of this chapter can be found in the Campanella's files *Mathtools (12/10/2011)* and *appnew (10/12/2011)*.

M. Campanella et al., *Interpretative Aspects of Quantum Mechanics*, UNIPA Springer Series,
https://doi.org/10.1007/978-3-030-44207-1

However, sometimes this notation is ambiguous. Indeed, in this notation, it is understood that the fundamental space is H, while the space H^* of the bras (i.e., the dual space of the linear functionals on H) is a derived construction. This circumstance is translated in the formalism using the ket notation $|x\rangle$ for the elements of H and the bra notation $\langle x|$ for the elements of H^*. But it will be essential in several points of our analysis to treat H and H^* on the same footing. This possibility is essentially a consequence of the fact that H and its double dual H^{**} can be identified. We can consider a bra of H (i.e., a linear functional on H) as a ket of H^* (and in this perspective we change the role of H^* from derived to fundamental) and consequently a bra of H^* (i.e., a linear functional on H^*, which is a derived construction if we regard H^* as fundamental) is an element of H^{**}, i.e., an element of H. In this way, the kets of H become a derived construction if H^* is regarded as fundamental.

All this is reflected into the notation in the following way. First, if H is considered fundamental, an element of it will be denoted $|x\rangle_H$. The corresponding bra (that is, the functional on H) will be denoted ${}_H\langle x|$. The same bra of H must be written as a ket of H^* if the latter is considered as fundamental (that is, as an element of the dual space H^* considered as a linear space in its own). We therefore have ${}_H\langle x| = |x\rangle_{H^*}$; similarly, changing H in H^*, we get ${}_{H^*}\langle x| = |x\rangle_H$. Furthermore we have $|x\rangle_H^* = {}_H\langle x|$ and, denoting with a star on the left the action of conjugation on a bra of H, we have also ${}^*_H\langle x| = |x\rangle_H$. We conclude that, besides the rules

$$|x\rangle_H^* = {}_H\langle x| \quad \text{and} \quad {}^*_H\langle x| = |x\rangle_H,$$

the rules

$${}_H\langle x| = |x\rangle_{H^*} \quad \text{and} \quad {}_{H^*}\langle x| = |x\rangle_H$$

hold. The rules

$$|x\rangle_H^* = |x\rangle_{H^*} \quad \text{and} \quad {}^*_H\langle x| = {}_{H^*}\langle x|$$

follow. Changing in the latter H with H^*, the rules

$$|x\rangle_{H^*}^* = |x\rangle_H \quad \text{and} \quad {}^*_{H^*}\langle x| = {}_H\langle x|$$

are obtained.

Let $H \xrightarrow{\varphi} K$ be a linear mapping between Hilbert spaces. If $|k\rangle_K$ is the image of $|h\rangle_H$, we write $|k\rangle_K = \varphi|h\rangle_H$.

The transpose of φ is a linear mapping $K^* \xrightarrow{\varphi^d} H^*$. We can then write

$$\varphi^d|y\rangle_{K^*} = {}_K\langle y|\,\varphi.$$

Putting $\varphi = \varphi'^d$ we get $\varphi'|y\rangle_{K^*} = {}_K\langle y|\,\varphi'^d$ and putting $K = H^*$ and calling φ the arbitrary linear mapping φ, we have

$$\varphi|y\rangle_H = {}_{H^*}\langle y|\,\varphi^d.$$

We define the *adjoint* of φ as

$$\varphi^* = (\,)^* \varphi^d (\,)^*,$$

so that the action of φ^* on a ket $|y\rangle_K$ is given by

$$|y\rangle_K \xrightarrow{*} |y\rangle_{K^*} \xrightarrow{\varphi^d} \varphi^d |y\rangle_{K^*} = {}_K\langle y|\,\varphi \xrightarrow{*} ({}_K\langle y|\,\varphi)^* = \varphi^* |y\rangle_K.$$

As a consequence of the definition of the adjoint, we have that

$$\text{if} \quad {}_H\langle x| = {}_K\langle y|\,\varphi, \quad \text{then} \quad |x\rangle_H = \varphi^* |y\rangle_K.$$

From the definition of adjoint, we get also $\varphi^{**} = \varphi$.

Let $|y\rangle_K = \varphi|x\rangle_H$. Hence $|y\rangle_K = {}_{H^*}\langle x|\,\varphi^d$. Consequently,

$${}_H\langle x| \xrightarrow{*} {}_{H^*}\langle x| \mapsto {}_{H^*}\langle x|\,\varphi^d = |y\rangle_K \xrightarrow{*} {}_K\langle y|\,.$$

Hence ${}_K\langle y| = {}_K\langle x|\,\varphi^*$.

C.2 Free Vector Spaces

We recall that a free vector space over S is a mapping $S \xrightarrow{\sigma} X$ from S in a vector space X such that every mapping $S \xrightarrow{\tau} Z$ in a vector space Z uniquely factorizes as $\tau = \vartheta \circ \sigma$ where ϑ is linear. It is well known that every vector space is free over some subset of it. Each such subset is called a (Hamel) basis.

The *standard free space* over S is the set F of finite support mappings $S \to \mathbb{C}$ with pointwise addition and scalar multiplication, together with the mapping $S \xrightarrow{\sigma} F$ defined as

$$\sigma(s)\left(s'\right) = \delta_{ss'}.$$

The image E of σ is a basis of F which will be called the standard basis of F. The restriction of σ to E is a bijection. The element of E corresponding to $s \in S$ will be denoted e_s. When S is finite, the finite support condition is trivially verified.

Every free linear space over S can be regarded as a free linear space over the basis corresponding to S. Conversely, every linear space with a selected basis E can be regarded as a free linear space over E.

Let $\{X_\alpha\}$ be a family of (not necessarily distinct) linear spaces. A *direct sum* of the family is a family of linear mappings $X_\alpha \xrightarrow{\iota_\alpha} X$ such that every family of linear mappings $X_\alpha \xrightarrow{f_\alpha} Z$ is uniquely factorizable as $f_\alpha = f \circ \iota_\alpha$ through the linear mapping f.

If $\{\iota_\alpha\}$ and $\{\iota'_\alpha\}$ are direct sums of the same family, there is a unique isomorphism $X \xrightarrow{\iota} X'$ of linear spaces such that $\iota'_\alpha = \iota \circ \iota_\alpha$. It can be shown that the ι_α are injective.

A possible direct sum of the family $\{X_\alpha\}$ is obtained as follows. We call $\mathscr{A}$ the set of all indexes α and define X as the set of all mappings

$$\mathscr{A} \xrightarrow{x} \bigcup_{\alpha \in \mathscr{A}} X_\alpha$$

of finite support and such that $x_\alpha \in X_\alpha$. X is a linear space under elementwise addition and scalar multiplication. We put

$$\iota_\alpha(x_\alpha)_\beta = \delta_{\alpha\beta} x_\alpha.$$

It is easily verified that these positions satisfy the definition of a direct sum. The direct sum obtained with the above construction will be called the *standard direct sum*. The space X will be denoted $\oplus_\alpha X_\alpha$ and the injections ι_α will be understood.

C.2.1 Finite-Dimensional Free Vector Spaces

We will consider here only finite-dimensional spaces.

Suppose that $\{S_\alpha\}$ is a family of finite sets with the same cardinality and that to each pair (β, α) a mapping $S_\beta \times S_\alpha \xrightarrow{t^{(\beta\alpha)}} \mathbb{C}$ is associated satisfying the following conditions:

$$\begin{aligned} &\text{a)}\ t^{(\alpha\alpha)}_{s's} = \delta_{s's}, \\ &\text{b)}\ \sum_{s''} t^{(\gamma\beta)}_{s's''} t^{(\beta\alpha)}_{s''s} = t^{(\gamma\alpha)}_{s's}. \end{aligned}$$

For each S_α, let F_α be (the codomain of) the standard free vector space over S_α and Σ their standard direct sum. Let Γ be the subspace of Σ generated by all the elements of the form

$$\iota_\beta\left(e^{(\beta)}_{s'}\right) - \sum_s t^{(\beta\alpha)}_{s's} \iota_\alpha\left(e^{(\alpha)}_s\right).$$

The quotient space Σ/Γ will be called the *gluing of the family* of the standard free spaces $\{F_\alpha\}$ through the family of *gluing mappings* $\{t^{(\beta\alpha)}\}$. This space will be denoted as

$$\mathscr{G}\left(\{F_\alpha\}, \{t^{(\beta\alpha)}\}\right).$$

We observe that the set $\left\{\iota_\beta(e^{(\beta)}_{s'})\right\}$ is a basis for Σ. Furthermore, for a fixed $\bar{\alpha}$, we define, for each $\beta \neq \bar{\alpha}$, the vectors

$$h_{s'}^{(\beta\bar{\alpha})} = \iota_\beta(e_{s'}^{(\beta)}) - \sum_s t_{s's}^{(\beta\bar{\alpha})} \iota_{\bar{\alpha}}(e_s^{(\bar{\alpha})}).$$

The vectors $h_{s'}^{(\beta\bar{\alpha})}$ together with the $\iota_{\bar{\alpha}}(e_{s'}^{(\bar{\alpha})})$ are a basis of Σ. Indeed, for

$$\beta \neq \bar{\alpha}, \quad \iota_\beta(e_{s'}^{(\beta)}) = h_{s'}^{(\beta\bar{\alpha})} + \sum_s t_{s's}^{(\beta\bar{\alpha})} \iota_{\bar{\alpha}}(e_s^{(\bar{\alpha})}).$$

This shows that the set $\left\{h_{s'}^{(\beta\bar{\alpha})}, \iota_{\bar{\alpha}}(e_{s'}^{(\bar{\alpha})})\right\}$ spans Σ. Furthermore this set is independent. Indeed, if

$$\sum_{\beta\neq\bar{\alpha}} \lambda_{s'}^{(\beta)} h_{s'}^{(\beta\bar{\alpha})} + \sum_{s'} \mu_{s'} \iota_{\bar{\alpha}}(e_{s'}^{(\bar{\alpha})}) = 0,$$

$$\sum_{\beta\neq\bar{\alpha}} \lambda_{s'}^{(\beta)} \iota_\beta(e_{s'}^{(\beta)}) - \sum_{\beta\neq\bar{\alpha}} \sum_s \lambda_{s'}^{(\beta)} t_{s's}^{(\beta\bar{\alpha})} \iota_{\bar{\alpha}}(e_s^{(\bar{\alpha})}) + \sum_{s'} \mu_{s'} \iota_{\bar{\alpha}}(e_{s'}^{(\bar{\alpha})}) = 0.$$

This entails $\lambda_{s'}^{(\beta)} = 0$ and hence $\mu_{s'} = 0$. We have

$$h_{s'}^{(\beta\alpha)} = \iota_\beta\left(e_{s'}^{(\beta)}\right) - \sum_s t_{s's}^{(\beta\alpha)} \iota_\alpha\left(e_s^{(\alpha)}\right) = h_{s'}^{(\beta\bar{\alpha})} + \sum_s t_{s's}^{(\beta\bar{\alpha})} \iota_{\bar{\alpha}}\left(e_s^{(\bar{\alpha})}\right) - \sum_s t_{s's}^{(\beta\alpha)} \iota_\alpha\left(e_s^{(\alpha)}\right).$$

But we can write

$$\sum_s t_{s's}^{(\beta\alpha)} \iota_\alpha\left(e_s^{(\alpha)}\right) = \sum_s t_{s's}^{(\beta\alpha)} h_s^{(\alpha\bar{\alpha})} + \sum_s t_{s's}^{(\beta\alpha)} \sum_{s''} t_{ss''}^{(\alpha\bar{\alpha})} \iota_{\bar{\alpha}}\left(e_{s''}^{(\bar{\alpha})}\right)$$

and

$$\sum_s t_{s's}^{(\beta\alpha)} \sum_{s''} t_{ss''}^{(\alpha\bar{\alpha})} \iota_{\bar{\alpha}}\left(e_{s''}^{(\bar{\alpha})}\right) = \sum_{s''} t_{s's''}^{(\beta\bar{\alpha})} \iota_{\bar{\alpha}}\left(e_{s''}^{(\bar{\alpha})}\right)$$

so that

$$h_{s'}^{(\beta\alpha)} = h_{s'}^{(\beta\bar{\alpha})} + \sum_s t_{s's}^{(\beta\bar{\alpha})} \iota_{\bar{\alpha}}\left(e_s^{(\bar{\alpha})}\right) - \sum_s t_{s's}^{(\beta\alpha)} h_s^{(\alpha\bar{\alpha})} - \sum_{s''} t_{s's''}^{(\beta\bar{\alpha})} \iota_{\bar{\alpha}}\left(e_{s''}^{(\bar{\alpha})}\right)$$

that is $h_{s'}^{(\beta\alpha)} = h_{s'}^{(\beta\bar{\alpha})} - \sum_s t_{s's}^{(\beta\alpha)} h_s^{(\alpha\bar{\alpha})}$.

We conclude that Γ admits $\left\{h_s^{(\beta\bar{\alpha})}\right\}$ as a basis. Hence, Σ is the internal direct sum of $\iota_{\bar{\alpha}}(F_{\bar{\alpha}})$ and Γ. Consequently, the quotient space $X = \Sigma/\Gamma$ has the basis

$$\left\{\bar{e}_s^{(\bar{\alpha})} = \iota_{\bar{\alpha}}\left(e_s^{(\bar{\alpha})}\right) + \Gamma\right\}.$$

As $\bar{\alpha}$ is arbitrary, every $\left\{\bar{e}_s^{(\alpha)} = \iota_\alpha\left(e_s^{(\alpha)}\right) + \Gamma\right\}$ is a basis of X. Consequently, every $x \in X$ has the unique expansion $x = \sum_s x_s^{(\alpha)} \bar{e}_s^{(\alpha)}$. The expansion of the same vector with respect to the basis $\left\{e_{s'}^{(\beta)}\right\}$ is $x = \sum_{s'} x_{s'}^{(\beta)} \bar{e}_{s'}^{(\beta)}$. As $\bar{e}_s^{(\alpha)} = \sum_{s'} t_{ss'}^{(\alpha\beta)} \bar{e}_{s'}^{(\beta)}$, we obtain $x_{s'}^{(\beta)} = \sum_s x_s^{(\alpha)} t_{ss'}^{(\alpha\beta)}$.

The above analysis leads to the following.

Theorem C.1 *if $\{F_\alpha\}$ is a family of free linear spaces over the sets S_α, and X the gluing of the family through the family $\left\{t^{(\beta\alpha)}\right\}$ of gluing mappings, each set*

$$\left\{\bar{e}_s^{(\alpha)} = \iota_\alpha \left(e_s^{(\alpha)}\right) + \Gamma\right\}$$

is a basis of X. If $x = \sum_s x_s^{(\alpha)} \bar{e}_s^{(\alpha)} = \sum_{s'} x_{s'}^{(\beta)} \bar{e}_{s'}^{(\beta)}$ then $x_{s'}^{(\beta)} = \sum_s x_s^{(\alpha)} t_{ss'}^{(\alpha\beta)}$.

Remark C.1 Let $x \in X$ be represented by $f \in \oplus_\alpha f_\alpha$. Given $\bar{\alpha}$, we have the expansion $x = \sum_s x_s^{(\bar{\alpha})} \bar{e}_s^{(\bar{\alpha})}$ that is $x = \sum_s x_s^{(\bar{\alpha})} e_s^{(\bar{\alpha})} + \Gamma$. This means that x can be represented by an element of $\iota_{\bar{\alpha}} (F_{\bar{\alpha}})$. This representation is unique (for given $\bar{\alpha}$), because $\Sigma = F_{\bar{\alpha}} + \Gamma$. This means also that every element of Σ is equivalent to a unique element of $\iota_{\bar{\alpha}} (F_{\bar{\alpha}})$ modulo Γ.

Suppose now that X is a finite-dimensional linear space. Let $\{S_\alpha\}$ a set of bases of X. For each S_α, we introduce the standard free linear space F_α over S_α. If $S_\alpha = \{f\}$ and $S_\beta = \{f'\}$ we have the expansion $f = \sum_{f'} t_{ff'}^{(\alpha\beta)} f'$.

We can then build the gluing $\hat{X}$ of the family $\{F_\alpha\}$ through the family of the gluing mappings $\left\{t_{ff'}^{(\alpha\beta)}\right\}$. Let $\hat{x} \in \hat{X}$. For a given α, there is a unique representative $\sum_f x_f^{(\alpha)} \iota_\alpha \left(e_f^{(\alpha)}\right)$ of $\hat{x}$ belonging to $\iota_\alpha (F_\alpha)$. The elements $e_f^{(\alpha)}$ of the standard basis of F_α are in bijection with the elements f of the basis S_α. We put $x = \sum_f x_f^{(\alpha)} f$. The element x does not depend on the choice of α. Indeed, if $\hat{x}$ is represented by $\sum_{f'} x_{f'}^{(\beta)} \iota_\beta \left(e_{f'}^{(\beta)}\right)$, then

$$x_{f'}^{(\beta)} = \sum_f x_f^{(\alpha)} t_{ff'}^{(\alpha\beta)}, \qquad x = \sum_f x_f^{(\alpha)} \sum_{f'} t_{ff'}^{(\alpha\beta)} f' = \sum_{f'} x_{f'}^{(\beta)} f'.$$

In this way, we get a linear mapping $\hat{X} \xrightarrow{j} X$. This mapping is clearly bijective, so that it is an isomorphism of linear spaces. In this way, we have proved the following theorem.

Theorem C.2 *If X is a finite-dimensional linear space and $\{S_\alpha\}$ is a set of bases of X, there is a canonical isomorphism j between the gluing $\hat{X}$ of the family $\{F_\alpha\}$ of the corresponding free linear spaces through the family of transition mappings* *between the bases and the space X. If* $\sum_{f \in S_\alpha} x_f^{(\alpha)} \iota_\alpha \left(e_f^{(\alpha)}\right)$ *is the unique representative of* $\hat{x}$ *belonging to* $\iota_\alpha(F_\alpha)$, $j(\hat{x}) = \sum_{f \in S_\alpha} x_f^{(\alpha)} f$.

We now consider a family $\left\{X_\alpha, \left\{e_s^{(\alpha)}\right\}\right\}$ of linear spaces, each equipped with a prescribed basis. It can be considered as a family of linear free spaces over the family of sets $\left\{\left\{e_s^{(\alpha)}\right\}\right\}$. If we prescribe a family $\left\{t^{(\beta\alpha)}\right\}$ of transition mappings, we get a gluing for the family. Let $\left\{\left\{\tilde{e}_s^{(a)}\right\}\right\}$ a new family of bases, and $\left\{\tilde{t}^{(ba)}\right\}$ a new family of transition mappings. We want to establish the necessary and sufficient conditions in order that the same gluing space is obtained. Fixing some $\bar{\alpha}$, Γ has a basis given by the set

$$\left\{h_{s'}^{(\beta\bar{\alpha})} = \iota_\beta\left(e_{s'}^{(\beta)}\right) - \sum_s t_{s's}^{(\beta\bar{\alpha})} \iota_{\bar{\alpha}}\left(e_s^{(\bar{\alpha})}\right)\right\}$$

with $\beta \neq \bar{\alpha}$, and similarly, fixing some $\bar{a}$, $\tilde{\Gamma}$ has a basis given by the set

$$\left\{\tilde{h}_{s'}^{(b\bar{a})} = \iota_b\left(\tilde{e}_{s'}^{(b)}\right) - \sum_s \tilde{t}_{s's}^{(b\bar{a})} \iota_{\bar{a}}\left(\tilde{e}_s^{(\bar{a})}\right)\right\}$$

with $b \neq \bar{a}$. But $\tilde{e}_{s'}^{(a)} = \sum_s u_{s's}^{(a\alpha)} e_s^{(\alpha)}$, so that

$$\tilde{h}_{s'}^{(b\bar{a})} = \iota_b\left(\sum_s u_{s's}^{(b\beta)} e_s^{(\beta)}\right) - \sum_{s''} \tilde{t}_{s's''}^{(b\bar{a})} \iota_{\bar{a}}\left(\sum_s u_{s''s}^{(\bar{a}\bar{\alpha})} e_s^{(\bar{\alpha})}\right).$$

We then get

$$\tilde{h}_{s'}^{(b\bar{a})} = \sum_s u_{s's}^{(b\beta)} h_s^{(\beta\bar{\alpha})} + \sum_s u_{s's}^{(b\beta)} \sum_{s''} t_{ss''}^{(\beta\bar{\alpha})} \iota_{\bar{\alpha}}\left(e_{s''}^{(\bar{\alpha})}\right) - \sum_{s''} \tilde{t}_{s's''}^{(b\bar{a})} \iota_{\bar{a}}\left(\sum_s u_{s''s}^{(\bar{a}\bar{\alpha})} e_s^{(\bar{\alpha})}\right).$$

The necessary and sufficient condition in order that $\tilde{\Gamma} = \Gamma$ is therefore

$$\sum_{s''} u_{s's''}^{(b\beta)} \sum_s t_{s''s}^{(\beta\bar{\alpha})} \iota_{\bar{\alpha}}\left(e_s^{(\bar{\alpha})}\right) - \sum_{s''} \tilde{t}_{s's''}^{(b\bar{a})} \iota_{\bar{a}}\left(\sum_s u_{s''s}^{(\bar{a}\bar{\alpha})} e_s^{(\bar{\alpha})}\right) = 0$$

that is

$$\sum_{s''} u_{s's''}^{(b\beta)} t_{s''s}^{(\beta\bar{\alpha})} - \sum_{s''} \tilde{t}_{s's''}^{(b\bar{a})} u_{s''s}^{(\bar{a}\bar{\alpha})} = 0.$$

Finally, remembering that $\bar{\alpha}$ and $\bar{a}$ are arbitrary (provided that they refer to the same space) we obtain

$$\tilde{t}^{(ba)}_{s's} = \sum_{s''s'''} u^{(b\beta)}_{s's''} t^{(\beta\alpha)}_{s''s'''} u^{(\alpha a)}_{s'''s}.$$

We have thus established the following theorem.

Theorem C.3 *Let $\{X_\alpha\}$ be a family of linear spaces with the same dimension. Let $\{E_\alpha\}, \left\{\tilde{E}_\alpha\right\}$ be two families of selected bases in the spaces X_α. Let $\left\{t^{(\beta\alpha)}\right\}, \left\{\tilde{t}^{(\beta\alpha)}\right\}$ two families of gluing mappings associated with the family of bases $\{E_\alpha\}$ and $\left\{\tilde{E}_\alpha\right\}$, respectively. Let $\tilde{e}^{(a)}_{s'} = \sum_s u^{(\alpha)}_{s's} e^{(\alpha)}_s$. The two families of gluing mappings yield the same gluing space if and only if*

$$\tilde{t}^{(\beta\alpha)}_{s's} = \sum_{s''s'''} u^{(\beta)}_{s's''} t^{(\beta\alpha)}_{s''s'''} u^{(\alpha)-1}_{s'''s}.$$

C.3 Tensor Products

We recall that a tensor product is a bilinear mapping

$$X \times Y \overset{\otimes}{\to} Z$$

such that every bilinear mapping $X \times Y \overset{\beta}{\to} V$ uniquely factorizes as $\beta = \varphi \circ \otimes$ through a linear mapping $Z \overset{\varphi}{\to} V$. The elements of the image of $\otimes$ are called *indecomposable* (or *separable*) elements. It is well known that, given X and Y, a tensor product with domain $X \times Y$ can be defined through the following construction. First the free linear space F over the set $X \times Y$ is introduced. Then the subspace Γ of F generated by all elements of the forms

$$(\lambda x, y) - \lambda (x, y), (x, \lambda y) - \lambda (x, y) \text{ and}$$

$$\left(x + x', y\right) - (x, y) - \left(x', y\right), \left(x, y + y'\right) - (x, y) - \left(x, y'\right).$$

If π is the canonical projection of F over F/Γ, we define $x \otimes y = \pi (x, y)$.

The above construction holds also for infinite-dimensional spaces. It is also well known that, given the spaces X and Y, a tensor product is defined "up to isomorphisms." This means that, if $X \times Y \overset{\otimes}{\to} Z$ and $X \times Y \overset{\otimes'}{\to} Z'$ are tensor products, there is a unique isomorphism of linear spaces $Z \overset{\eta}{\to} Z'$ such that $\otimes' = \eta \circ \otimes$.

When X and Y are finite-dimensional, another way of defining a tensor product is possible. We first observe that a canonical linear mapping $X \overset{\eta}{\to} X^{dd}$ from X to its double dual can be defined in the following way. If $\xi \in X^d$ and $x \in X$, we introduce the notation

$$\langle \xi, x \rangle = \xi (x)$$

Then for each x $\langle \xi, x \rangle$ depends linearly on ξ, so that an element $\eta (x)$ of X^{dd} arises. The dependence of $\eta (x)$ on x is linear, and hence η is a linear mapping from X to X^{dd}. In the above notation, η is uniquely defined by the equation $\langle \eta (x), \xi \rangle = \langle \xi, x \rangle$. When X is finite-dimensional, it is easily shown that η is an isomorphism.

We now introduce the linear space $Hom\left(X^d, Y\right)$. For each $(x, y) \in X \times Y$, we define an element $x \otimes y \in Hom\left(X^d, Y\right)$ through the position

$$(x \otimes y)(\xi) = \langle \eta (x), \xi \rangle \, y = \langle \xi, x \rangle \, y.$$

The mapping $\otimes$ is clearly bilinear. Furthermore, let $X \times Y \overset{\beta}{\to} V$ a bilinear mapping.

We introduce bases $\{e_\alpha\}_{\alpha \in \mathscr{A}}$ and $\{f_b\}_{b \in B}$ of X and Y, respectively. Then $\{e_\alpha \otimes f_b\}_{\alpha \in \mathscr{A}, b \in B}$ is a basis of $Hom\left(X^d, Y\right)$. Indeed $\{\eta (e_\alpha)\}_{\alpha \in \mathscr{A}}$ is a basis of $\left(X^d\right)^d$, so that the set $\{\bar{e}_{\alpha'}\}_{\alpha' \in \mathscr{A}}$ defined by $\langle \eta (e_\alpha), \bar{e}_{\alpha'} \rangle = \delta_{\alpha\alpha'}$ is a basis of X^d.

Furthermore, the elements of $Hom\left(X^d, Y\right)$ defined by $\varphi_{\alpha b} (\bar{e}_{\alpha'}) = \delta_{\alpha\alpha'} f_b$ form a basis. But $(e_\alpha \otimes f_b)(\bar{e}_{\alpha'}) = \delta_{\alpha\alpha'} f_b$, so that $e_\alpha \otimes f_b = \varphi_{\alpha b}$. Now if $\beta = \varphi \circ \otimes$, we get $\beta (e_\alpha, f_b) = \varphi (e_\alpha \otimes f_b)$, so that φ must be the unique linear mapping that takes the values $\beta (e_\alpha, f_b)$ on the elements of the basis $\{e_\alpha \otimes f_b\}_{\alpha \in \mathscr{A}, b \in B}$.

Conversely, if φ is defined in this way, we get $\beta = \varphi \circ \otimes$ by bilinearity. Thus $\otimes$ is a tensor product. In the case of finite-dimensional spaces X and Y, the mapping $\otimes$ defined above will be called the *canonical tensor product* of X and Y. For this tensor product, the separable elements are the elements of $Hom\left(X^d, Y\right)$ of rank at most one. Indeed, if either x or y is zero, $x \otimes y$ is of rank zero. If both x and y are nonzero, then $x \otimes y$ is of rank one. Conversely, the zero homomorphism is the image of $0 \otimes 0$ and a homomorphism of rank one must have the form $\varphi (\xi) = \vartheta (\xi) \, y$ for some fixed $y \in Y$. ϑ must be a linear functional on X^d and hence an element of X^{dd}. Hence $\vartheta = \eta (x)$ and $\varphi (\xi) = \langle \eta (x), \xi \rangle \, y$, so that $\varphi = x \otimes y$. We can summarize the above results in the following theorem.

Theorem C.4 *If X and Y are finite-dimensional linear spaces, the mapping*

$$X \times Y \overset{\otimes}{\to} Hom\left(X^d, Y\right)$$

defined by the position

$$(x \otimes y)(\xi) = \langle \xi, x \rangle \, y$$

is a tensor product. The separable elements are all the elements of $Hom\left(X^d, Y\right)$ whose rank is not greater than one; each pair of bases $\{e_\alpha\}_{\alpha \in \mathscr{A}}$ and $\{f_b\}_{b \in B}$ of X and Y, respectively, defines a basis

$$\{e_\alpha \otimes f_b\}_{\alpha \in \mathscr{A}, b \in B} \quad of \quad Hom\left(X^d, Y\right)$$

formed uniquely with separable elements.

Let $X \times Y \overset{\otimes}{\to} Z$ and $X' \times Y' \overset{\otimes'}{\to} Z$ be two tensor products. We say that $\otimes$ and $\otimes'$ are *equivalent* if there are linear space isomorphisms $X' \overset{\varphi}{\to} X$ and $Y' \overset{\vartheta}{\to} Y$ such that

$$\otimes' = \otimes \circ (\varphi \times \vartheta) .$$

We say that a *structure of tensor product* is defined in Z when a class of equivalence of tensor products whose codomain is Z is specified. Let $X \times Y \overset{\otimes}{\to} Z$ be a representative of the class of equivalence. Let E and F be bases of X and Y, respectively. Then $E \otimes F$ (i.e., the set of all elements of the form $e \otimes f$ with $e \in E,\ f \in F$) is a basis of Z. There is a bijection $E \times F \overset{\eta}{\to} E \otimes F$ defined by $\eta\,(e, f) = e \otimes f$. The mapping η is obviously surjective. The injectivity follows immediately from the fact that $e' \otimes f' = e \otimes f$ entails $e' = \lambda e$ and from the fact that the elements of E are linearly independent. If π_E and π_F denote the natural projections of $E \times F$, we can define two equivalence relations on $E \otimes F$ as follows:

$$u \sim v \quad \Leftrightarrow \quad \pi_E \eta^{-1}(u) = \pi_E \eta^{-1}(v) ,$$

$$u \approx v \quad \Leftrightarrow \quad \pi_F \eta^{-1}(u) = \pi_F \eta^{-1}(v) .$$

Let $\pi_\sim$ and $\pi_\approx$ denote the natural projections of $\sim$, $\approx$ and $C_\sim$, $C_\approx$ the corresponding images. The intersection of any element of $C_\sim$ with any element of $C_\approx$ is a *singleton*.

We say that two equivalence relations on the same set are *transversal* if every intersection of a class of equivalence c_1 of one of them with a class of equivalence c_2 of the other is a singleton. The element defined by this singleton will be denoted $c_1 \wedge c_2$. There are a unique bijection $C_\sim \overset{\alpha}{\to} E$ such that $\alpha \circ \pi_\sim = \pi_E \circ \eta^{-1}$ and a unique bijection $C_\approx \overset{\beta}{\to} F$ such that $\beta \circ \pi_\approx = \pi_F \circ \eta^{-1}$.

We denote $\hat{X}$ the free linear space over $C_\sim$ and $\hat{Y}$ the free linear space over $C_\approx$. For the sake of notational simplicity, we will identify $C_\sim$ and $C_\approx$ with their images in $\hat{X}$ and $\hat{Y}$, respectively.

We now define $\hat{\otimes}$ as the unique bilinear extension of $c_1 \wedge c_2$ to the whole $\hat{X} \times \hat{Y}$, thus obtaining a tensor product with codomain Z.

If we now define φ as the linear extension of α to the whole $\hat{X}$ and ϑ as the linear extension of β to the whole $\hat{Y}$, we obtain $\hat{\otimes} = \otimes \circ (\varphi \times \vartheta)$.

Conversely, let B be a basis of Z and $(\sim, \approx)$ a pair of transversal equivalence relations on it. If we define the sets $C_\sim$ and $C_\approx$ of equivalence classes, the corresponding free spaces $\hat{X}$ and $\hat{Y}$ and $\hat{\otimes}$ as the bilinear extension of $c_1 \wedge c_2$, we get a tensor product with codomain Z.

We call this product the *standard tensor product* associated with $(B, \sim, \approx)$.

Definition C.1 Let $(B, \sim, \approx)$ be a basis of the linear space Z together with a pair of transversal equivalence relations on it. If $C_\sim$ and $C_\approx$ are the corresponding sets of equivalence classes, the unique tensor product from the Cartesian product of the free spaces $\hat{X}$ and $\hat{Y}$ over $C_\sim$ and $C_\approx$, respectively, into Z defined by the mapping $c_1 \wedge c_2$ is called the standard tensor product associated with $(B, \sim, \approx)$.

The above discussion can be summarized in the following theorem.

Theorem C.5 *For every tensor product $\otimes$ whose codomain is the linear space Z there are a basis B of Z and a pair $(\sim, \approx)$ of transversal equivalence relations on B such that $\otimes$ is equivalent to the standard tensor product $\hat{\otimes}$ associated with $(B, \sim, \approx)$. Furthermore, for every $(B, \sim, \approx)$ the associated standard tensor product has Z as its codomain and thus defines a structure of tensor product on it.*

On the basis of the above theorem we conclude that for every structure of tensor product on Z there is a representative consisting in a standard tensor product associated with some $(B, \sim, \approx)$.

Nevertheless, different $(B, \sim, \approx)$ can give rise to the same structure of tensor product on Z. Hence, we are led to investigate the necessary and sufficient conditions to be satisfied in order that $(B, \sim, \approx)$ and $(B', \sim', \approx')$ define equivalent standard tensor products. The equivalence entails the existence of $\hat{X}' \xrightarrow{\varphi} \hat{X}$ and $\hat{Y}' \xrightarrow{\vartheta} \hat{Y}$ such that $\hat{\otimes}' = \hat{\otimes} \circ (\varphi \times \vartheta)$. If $C'_\sim$ and $C'_\approx$ are the sets of equivalence classes associated with $\sim'$ and $\approx'$, $\varphi\left(C'_\sim\right)$ and $\vartheta\left(C'_\approx\right)$ are bases of $\hat{X}$ and $\hat{Y}$. Hence, for $c'_\sim \in C'_\sim$ we can write $\varphi\left(c'_\sim\right) = \sum_{c_\sim \in C_\sim} a\left(c'_\sim, c_\sim\right) c_\sim$. Similarly, $\vartheta\left(c'_\approx\right) = \sum_{c_\approx \in C_\approx} b\left(c'_\approx, c_\approx\right) c_\approx$. Hence, we get

$$c'_\sim \wedge c'_\approx = \sum_{c_\sim \in C_\sim} \sum_{c_\approx \in C_\approx} a\left(c'_\sim, c_\sim\right) b\left(c'_\approx, c_\approx\right) c_\sim \wedge c_\approx.$$

But, given $(B, \sim, \approx)$ and $(B', \sim', \approx')$, we have the unique expansion

$$c'_\sim \wedge c'_\approx = \sum_{c_\sim \in C_\sim} \sum_{c_\approx \in C_\approx} t\left(c'_\sim, c'_\approx, c_\sim, c_\approx\right) c_\sim \wedge c_\approx \tag{1.1}$$

so that there must be mappings a and b such that $t\left(c'_\sim, c'_\approx, c_\sim, c_\approx\right) = a\left(c'_\sim, c_\sim\right) b\left(c'_\approx, c_\approx\right)$. We observe that the mappings a and b are nonsingular, in the sense that the set of elements $\sum_{c_\sim \in C_\sim} a\left(c'_\sim, c_\sim\right) c_\sim$ and $\sum_{c_\approx \in C_\approx} b\left(c'_\approx, c_\approx\right) c_\approx$ are bases of $\hat{X}$ and $\hat{Y}$, respectively.

Conversely, if there are nonsingular mappings a and b such that $t\left(c'_\sim, c'_\approx, c_\sim, c_\approx\right) = a\left(c'_\sim, c_\sim\right) b\left(c'_\approx, c_\approx\right)$, putting

$$\varphi\left(c'_\sim\right) = \sum_{c_\sim \in C_\sim} a\left(c'_\sim, c_\sim\right) c_\sim$$

and

$$\vartheta\left(c'_{\approx}\right)=\sum_{c_{\approx}\in C_{\approx}} b\left(c'_{\approx}, c_{\approx}\right) c_{\approx}$$

we can uniquely extend by linearity φ and ϑ getting $\hat{\otimes}'=\hat{\otimes}\circ(\varphi\times\vartheta)$.

This result can be formulated in a more abstract form. Each $\tau \triangleq (B, \sim, \approx)$ defines a tensor product

$$\hat{X}_\tau \times \hat{Y}_\tau \xrightarrow{\hat{\otimes}_\tau} Z.$$

Each basis B of Z defines an isomorphism j_τ from the free space Z_τ over B to the space Z. We define

$$\tilde{\otimes}_\tau = j_\tau^{-1}\circ\hat{\otimes}_\tau$$

whose domain is the Cartesian product of the free spaces over $C_\sim$ and $C_\approx$ and whose codomain is the free space Z_τ over B_τ.

Equation (1.1) represents the isomorphism ${j_\tau}^{-1}\circ j_{\tau'}$. The existence of the mappings a and b means that ${j_\tau}^{-1}\circ j_{\tau'}$ is a tensor product.

In this way, a small category is defined whose objects are the free spaces $X^{(\tau)}$ over $C_\sim^{(\tau)}$. The set of morphisms $\Phi^{(\tau'\tau)}=Mor\left(X^{(\tau)}, X^{(\tau')}\right)$ is defined as the set of left factors of the factorization of ${j_{\tau'}}^{-1}\circ j_\tau$ as a tensor product.

The above definitions furnish a subcategory K of the whole category of linear spaces whose set of objects is the family $\{X_\tau\}$. Indeed, $\Phi^{(\tau\tau)}$ is the set of left factors of id_{Z_τ}, so that $id_{X_\tau}\in\Phi^{(\tau\tau)}$. Furthermore, if $\varphi'\in\Phi^{(\tau''\tau')}$ and $\varphi\in\Phi^{(\tau'\tau)}$, we have $\varphi'\otimes\vartheta'={j_{\tau''}}^{-1}\circ j_{\tau'}$ for some ϑ' and $\varphi\otimes\vartheta={j_{\tau'}}^{-1}\circ j_\tau$ for some ϑ, so that

$$\varphi'\circ\varphi\otimes\vartheta'\circ\vartheta={j_{\tau''}}^{-1}\circ j_\tau \text{ and } \varphi'\circ\varphi\in\Phi^{(\tau''\tau)}.$$

An immediate consequence of this definition is that, if φ and φ' belong to $\Phi^{(\tau'\tau)}$, then $\varphi'=\lambda\varphi$. In particular, $\Phi^{(\tau\tau)}$ consists of all scalar nonzero multiples of id_{X_τ}.

We now prove the existence of a subcategory S of K with the same objects and containing a single morphism for each pair of objects. We start choosing a $\varphi^{(\tau'\tau)}$ for each $\Phi^{(\tau'\tau)}$. The morphism $\varphi^{(\tau'\tau)}\circ\varphi^{(\tau\tau')}=\lambda^{(\tau'\tau)}id_{X_{\tau'}}$. Similarly $\varphi^{(\tau\tau')}\circ\varphi^{(\tau'\tau)}=\lambda^{(\tau\tau')}id_{X_\tau}$. Multiplying the last on the left by $\varphi^{(\tau'\tau)}$ we get $\lambda^{(\tau'\tau)}\varphi^{(\tau'\tau)}=\lambda^{(\tau\tau')}\varphi^{(\tau'\tau)}$, whence (observing that $\varphi^{(\tau'\tau)}$ is nonzero) $\lambda^{(\tau'\tau)}=\lambda^{(\tau\tau')}$. Defining

$$\bar{\varphi}^{(\tau'\tau)}=\left(\lambda^{(\tau'\tau)}\right)^{-\frac{1}{2}}\varphi^{(\tau'\tau)},$$

$\bar{\varphi}^{(\tau'\tau)}\in\Phi^{(\tau'\tau)}$ and we get $\bar{\varphi}^{(\tau'\tau)}\circ\bar{\varphi}^{(\tau\tau')}=id_{X_{\tau'}}$. This means that $\bar{\varphi}^{(\tau'\tau)}$ and $\bar{\varphi}^{(\tau\tau')}$ are reciprocally inverse isomorphisms. We have

$$\bar{\varphi}^{(\tau''\tau')} \circ \bar{\varphi}^{(\tau'\tau)} = \lambda^{(\tau''\tau'\tau)} \bar{\varphi}^{(\tau''\tau)}.$$

By cyclic permutation we get $\bar{\varphi}^{(\tau'\tau)} \circ \bar{\varphi}^{(\tau\tau'')} = \lambda^{(\tau'\tau\tau'')} \bar{\varphi}^{(\tau'\tau'')}$. Multiplying on the left by $\bar{\varphi}^{(\tau''\tau')}$ we obtain $\lambda^{(\tau''\tau'\tau)} = \lambda^{(\tau'\tau\tau'')}$. Furthermore, taking the inverse morphisms, we get

$$\lambda^{\tau''\tau\tau'} = \left(\lambda^{\tau''\tau'\tau}\right)^{-1}.$$

If we put $\bar{\varphi}^{(\tau''\tau')} = \mu^{(\tau''\tau')} \tilde{\varphi}^{(\tau''\tau')}$ and so on for the other morphisms, we can get $\tilde{\varphi}^{(\tau''\tau')} \circ \tilde{\varphi}^{(\tau'\tau)} \circ \tilde{\varphi}^{(\tau\tau'')} = id_{X_{\tau''}}$ provided that $\mu^{(\tau''\tau')} \mu^{(\tau'\tau)} \mu^{(\tau\tau'')} = \lambda^{(\tau''\tau'\tau)}$.

We now show that these equations possess a solution. Let us fix $\tau'' = \tau_0$. We put $\mu^{(\tau'\tau)} = \lambda^{(\tau_0\tau'\tau)}$. Substituting we get

$$\lambda^{(\tau_0\tau''\tau')} \lambda^{(\tau_0\tau'\tau)} \lambda^{(\tau_0\tau\tau'')} \left(\lambda^{(\tau''\tau'\tau)}\right)^{-1} = 1$$

which can be written as

$$\lambda^{(\tau''\tau'\tau)} \left(\lambda^{(\tau_0\tau'\tau)}\right)^{-1} \lambda^{(\tau_0\tau''\tau)} \left(\lambda^{(\tau_0\tau''\tau')}\right)^{-1} = 1.$$

The latter equation is satisfied. Indeed, it can be shown with some algebra that the latter equation is a consequence of the definition of the quantities λ in terms of the mappings $\bar{\varphi}$.

Our final conclusion is that there is a subcategory S of K with the same objects and containing a single morphism for each pair of objects. Such a subcategory is far from being unique. If $\{\tilde{\varphi}^{(\tau\tau')}\}$ and $\{\tilde{\varphi}'^{(\tau\tau')}\}$ are the families of morphisms of S and S', respectively, we must have

$$\tilde{\varphi'}^{(\tau\tau')} = \lambda^{(\tau\tau')} \tilde{\varphi}^{(\tau\tau')}.$$

The conditions $\tilde{\varphi}^{(\tau\tau')} \circ \tilde{\varphi}^{(\tau'\tau)} = id_\tau, \tilde{\varphi}^{(\tau''\tau')} \circ \tilde{\varphi}^{(\tau'\tau)} \circ \tilde{\varphi}^{(\tau\tau'')} = id_{\tau''}$ together with the corresponding conditions for the $\tilde{\varphi}'$ entail the equations

$$\lambda^{(\tau\tau')} \lambda^{(\tau'\tau)} = 1 \text{ and } \lambda^{(\tau''\tau')} \lambda^{(\tau'\tau)} \lambda^{(\tau\tau'')} = 1.$$

The most general solution of these equations is given by $\lambda^{(\tau'\tau)} = \mu^{(\tau')} \left(\mu^{(\tau)}\right)^{-1}$ for arbitrary values of the quantities μ. Indeed, the above positions satisfy identically the equations. Conversely, if we fix $\tau'' = \tau_0$, we get $\lambda^{(\tau'\tau)} = \lambda^{(\tau'\tau_0)} \left(\lambda^{(\tau\tau_0)}\right)^{-1}$. The quantities μ are defined up to an arbitrary (nonzero) common factor.

We now fix a subcategory S and introduce $\hat{X} = \oplus_\tau \hat{X}^{(\tau)}$ as the standard direct sum of the family of free spaces $\hat{X}^{(\tau)}$. We define in it the subspace Γ generated by all the elements of the form $e^{(\tau)} - \varphi^{(\tau'\tau)}\left(e^{(\tau)}\right)$ with $\varphi^{(\tau'\tau)} \in S$ for arbitrary τ, τ' and for arbitrary $e^{(\tau)}$ belonging to the standard basis of $\hat{X}^{(\tau)}$. This subspace is independent of the choice of S.

C.3.1 Some Properties of Tensor Products Between Linear Spaces

Let S be a set. Let $(\sim, \approx)$ be a pair of equivalence relations on S. We say that they are a transversal pair of equivalence relations (t.p. for brevity) if the intersection of two arbitrary equivalence classes of $\sim$ and $\approx$, respectively, is a singleton.

If p_1 and p_2 are the canonical projections associated with $\sim$ and $\approx$, respectively, and S_1, S_2 are their images, the pair (p_1, p_2) is a direct product of S_1 and S_2. Indeed, the Cartesian product $S_1 \times S_2$ together with its natural projections π_1, π_2 is a direct product. The unique mapping $S \xrightarrow{j} S_1 \times S_2$ defined by p_1, p_2 is a bijection. Furthermore, $\pi_i \circ j = p_i$, so that (p_1, p_2) is a direct product. A pair (p_1, p_2) obtained in the way described above will be called an internal direct product associated with S.

Theorem C.6 *If L is a subspace of $Z \otimes W$ consisting uniquely of indecomposable elements, it has the form $z \otimes L'$ where $z \in Z$ and L' is a subspace of W or $L' \otimes w$ where $w \in W$ and L' is a subspace of Z. The subspace L' is uniquely defined by L while z (or w) is defined up to a factor. The position $l' \mapsto z \otimes l'$ (or $l' \mapsto l' \otimes w$) defines a linear space isomorphism between L' and L.*

Proof If L is zero or one-dimensional, the thesis is obvious. Otherwise, let $l_1 = z_1 \otimes w_1$ and $l_2 = z_2 \otimes w_2$ be linearly independent elements of L. Then $l = \alpha l_1 + \beta l_2 = \alpha z_1 \otimes w_1 + \beta z_2 \otimes w_2$ belongs to L. If z_1, z_2 and w_1, w_2 are linearly independent, we can include z_1, z_2 in a basis $\{z_\alpha\}$ of Z and w_1, w_2 in a basis $\{w_\mu\}$ of W. Therefore, the expansion coefficients $l_{\alpha\mu}$ of l are $l_{11} = \alpha$ $l_{12} = 0$ $l_{21} = 0$ $l_{22} = \beta$, while all the remaining coefficients are zero. We conclude that L contains elements which are not indecomposable. From the contradiction, we conclude that either z_1, z_2 or w_1, w_2 must be collinear. Suppose, for instance, that z_1 and z_2 are collinear. If we rescale w_1 and w_2 we have $l_1 = z \otimes w_1, l_2 = z \otimes w_2$.

Let $l \in L$ be collinear with l_1 o r l_2. Then it can be represented as $l = z \otimes w$. If l is not collinear with l_1 and with l_2, suppose that $l = z' \otimes w$ with z' not collinear with z. Then w must be collinear with both w_1 and w_2. But this is impossible, because w_1 and w_2 are independent. Hence $l = z \otimes w$. We conclude that there is $z \in Z$ such that every $l \in L$ can be represented as $l = z \otimes w$.

The position $w \mapsto z \otimes w$ defines an injection W into $Z \otimes W$. Hence, for every $l \in L$ there is a unique w such that $l = z \otimes w$. The set of such elements w is a subspace L' of W, so that $L = z \otimes L'$ and the position $l' \mapsto z \otimes l'$ defines a linear space isomorphism between L' and L. If $z \otimes L' = \bar{z} \otimes \bar{L}'$, then $L' = \bar{L}'$ and $z = \lambda \bar{z}$. Indeed, if $x = z \otimes l' = \bar{z} \otimes \bar{l}', \bar{z} = \lambda z$ and $\bar{l}' = \lambda^{-1} l'$. Had we supposed that w_1 and w_2 are collinear, we would have concluded that $L = L' \otimes w$ with $w \in W$ and $L' \subseteq Z$.

C.3.2 Linear Mappings Between Tensor Products

Let X, Y, Z, W be linear spaces. We want to find the most general form of a linear mapping T from $X \otimes Y$ to $Z \otimes W$ which sends indecomposable elements into indecomposable elements.

The mapping T is uniquely defined by its values on the set of indecomposable elements of $X \otimes Y$. For each y we define

$$T_y(x) = T(x \otimes y).$$

If T transforms indecomposable elements into indecomposable elements, the image of the linear mapping T_y is a subset L of the set S of the indecomposable elements of $Z \otimes W$ and a linear subspace of $Z \otimes W$. Owing to Theorem C.6, the image of T_y has the form $L'_y \otimes w$ or $z \otimes L'_y$.

Let us consider the case when there are y and y' such that the image of T_y has the form $L'_y \otimes w$ and the image of $T_{y'}$ has the form $z' \otimes L'_{y'}$. Hence, we can define $\hat{T}_y$ and $\hat{T}_{y'}$ such that $T_y(x) = \hat{T}_y(x) \otimes w$ and $T_{y'}(x) = z' \otimes \hat{T}_{y'}(x)$. As

$$T_y(x) + T_{y'}(x) = T\left(x \otimes \left(y + y'\right)\right),$$

$\hat{T}_y(x) \otimes w + z' \otimes \hat{T}_{y'}(x)$ is indecomposable for every x. Reasoning as in the proof of theorem LINS, if X_1 is the linear subspace of X such that $\hat{T}_y(x)$ is collinear with z' and X_2 is the linear subspace of X such that $\hat{T}_{y'}(x)$ is collinear with w, we must have $X_1 \cup X_2 = X$. This is possible only if either X_1 or X_2 is the whole X. Indeed, consider the quotient space $X/(X_1 \cap X_2)$. If u does not belong to $X_1/(X_1 \cap X_2)$, it is represented by an element not belonging to X_1, so that it is represented by an element of X_2. Hence $X/(X_1 \cap X_2)$ is the point-set union of the spaces $X_1/(X_1 \cap X_2)$ and $X_2/(X_1 \cap X_2)$. As there is an inclusion preserving bijection between the subspaces of a quotient and the subspaces of the "numerator" containing the "denominator," the space $(X_1/(X_1 \cap X_2)) \cap (X_2/(X_1 \cap X_2))$ corresponds to $X_1 \cap X_2$ and thus it is zero. Hence $X/(X_1 \cap X_2)$ is both the union and the direct sum of two of its subspaces. This is possible only if at least one of these spaces is zero, because otherwise the sum of two nonzero elements belonging to different subspaces would belong to neither subspace. We conclude that either $X_1 \subseteq X_2$ or $X_2 \subseteq X_1$, so that either $X_1 = X$ or $X_2 = X$.

We conclude that either the image of $\hat{T}_y$ is the space generated by z' or the image of $\hat{T}_{y'}$ is the space generated by w. Consequently, either the image of T_y or the image of $T_{y'}$ is one-dimensional. Hence, if there are y and y' such that the image of T_y has the form $L'_y \otimes w$ and the image of $T_{y'}$ has the form $z' \otimes L'_{y'}$, there is $\bar{y}$ such that the image of $T_{\bar{y}}$ is one-dimensional, so that it is generated by an element of the form $\bar{z} \otimes \bar{w}$.

Consider now an arbitrary y. The mapping T_y is either of the form $\hat{T}_y(x) \otimes w$ or of the form $z \otimes \hat{T}_y(x)$. As $T_y(x) + T_{\bar{y}}(x)$ is indecomposable, in the first case either

w is collinear with $\bar{w}$ or the image of $\hat{T}_y$ is generated by $\bar{z}$. In the first of the latter alternatives, by suitable rescaling of $\hat{T}_y$ for each y we can write $T_y(x) = A_y(x) \otimes \bar{w}$, where A_y is a linear mapping. The dependence of $A_y(x)$ on y is linear, so that there is a linear mapping $X \otimes Y \xrightarrow{A} Z$ such that $A_y(x) = A(x \otimes y)$.

We conclude that in the first alternative $T(x \otimes y) = A(x \otimes y) \otimes \bar{w}$ for some $\bar{w} \in W$and some linear mapping A from $X \otimes Y$ to Z.

Always in the first case, but in the second alternative, $\hat{T}_y(x) = \lambda_y(x)\bar{z}$ so that we can write $T_y(x) = \bar{z} \otimes \lambda_y(x) w$. The element $\lambda_y(x) w$ must depend linearly on x, so that there is a linear mapping B_y such that $T_y(x) = \bar{z} \otimes B_y x$. The mapping which associates to each pair (x, y) of the element $B_y x$ is bilinear. Hence, there is a unique linear mapping $X \otimes Y \xrightarrow{B} W$ such that $B_y x = B(x \otimes y)$. We conclude that in the second alternative $T(x \otimes y) = \bar{z} \otimes B(x \otimes y)$ for some $\bar{z} \in Z$ and some linear mapping B from $X \otimes Y$ to W. Similar conclusions are obtained in the case when the mapping T_y has the form $z \otimes \hat{T}_y(x)$.

The result of the above analysis is that, when there are y and y' such that the image of T_y has the form $L'_y \otimes w$ and the image of $T_{y'}$ has the form $z' \otimes L'_{y'}$, either there are $\bar{w} \in W$ and $X \otimes Y \xrightarrow{A} Z$ such that $T(x \otimes y) = A(x \otimes y) \otimes \bar{w}$ or there are $\bar{z} \in Z$ and $X \otimes Y \xrightarrow{B} W$ such that $T(x \otimes y) = \bar{z} \otimes B(x \otimes y)$. By linearity, we can say that in the first case there are $\bar{w} \in W$ and $X \otimes Y \xrightarrow{A} Z$ such that $T(u) = A(u) \otimes \bar{w}$ and in the second case there are $\bar{z} \in Z$ and $X \otimes Y \xrightarrow{B} W$ such that $T(u) = \bar{z} \otimes B(u)$. We can characterize both cases by saying that T is the composition of a linear mapping from $X \otimes Y$ to one of the factors of $Z \otimes W$ with the natural injection of the factor in $Z \otimes W$ associated with a given vector of the other. A linear mapping from $X \otimes Y$ to $Z \otimes W$ which preserves the indecomposable elements and with the above property will be called of the first kind. Otherwise, it will be called of the second kind. The above analysis shows that, if there are y and y' such that the image of T_y has the form $L'_y \otimes w$ and the image of $T_{y'}$ has the form $z' \otimes L'_{y'}$, T must be of the first kind.

Suppose now that T is of the second kind. As a consequence, either the image of T_y has the form $L'_y \otimes w$ for all y or it has the form $z \otimes L'_y$ for all y. Let us consider the first case. Let us fix an element $\bar{y}$. The linear mapping $T_{\bar{y}}$ induces a linear mapping $T'_{\bar{y}}$ whose image is $L'_{\bar{y}}$, so that $T_{\bar{y}}(x) = T'_{\bar{y}}(x) \otimes \bar{w}$ Similarly, for an arbitrary y, $T_y(x) = T'_y(x) \otimes w$. We have

$$T_{\bar{y}}(x) + T_y(x) = T'_{\bar{y}}(x) \otimes \bar{w} + T'_y(x) \otimes w.$$

But $T_{\bar{y}}(x) + T_y(x)$ is indecomposable, so that either $T'_{\bar{y}}(x)$ and $T'_y(x)$ are collinear or $\bar{w}$ and w are collinear. Let us fix a value $\bar{x}$ of x. Put $\bar{z} = T'_{\bar{y}}(\bar{x})$. As $T_y(\bar{x})$ is linear with respect to y, $T'_y(\bar{x})$ enjoys the same property, so that the subset Y_1 of Y whose elements y are such that $T'_y(\bar{x})$ is collinear with $\bar{z}$ is a linear space. On the other hand, the subset Y_2 of Y whose elements y are such that w is collinear with $\bar{w}$ is a linear space. Indeed, let $T_{y_1}(x) = T'_{y_1}(x) \otimes \lambda_1\bar{w}$ and $T_{y_2}(x) = T'_{y_2}(x) \otimes \lambda_2\bar{w}$. Then

$$T_{\alpha y_1+\beta y_2}(x) = \alpha T'_{y_1}(x) \otimes \lambda_1 \bar{w} + \beta T'_{y_2}(x) \otimes \lambda_2 \bar{w} = \left(\alpha\lambda_1 T'_{y_1}(x) + \beta\lambda_2 T'_{y_2}(x)\right) \otimes \bar{w}.$$

We conclude that either w is collinear with $\bar{w}$ for all y or $T'_y(\bar{x})$ is collinear with $\bar{z}$ for all y. In the first alternative, we get $T_y(x) = T'_y(x) \otimes \lambda_y \bar{w}$. For each y, we can rescale T'_y, so that we get $T_y(x) = T'_y(x) \otimes \bar{w}$. T'_y depends linearly on y, so that there is a linear mapping $X \otimes Y \xrightarrow{A} Z$ such that $T(x \otimes y) = A(x \otimes y) \otimes \bar{w}$. But in this case T is of the first kind. Hence, the only alternative for a mapping of the second kind is that $T'_y(\bar{x})$ is collinear with $\bar{z}$ for all y. In this alternative, we can write $T'_y(\bar{x}) = \lambda_y(\bar{x})\,\bar{z}$.

Remembering the definition of $\bar{z}$, we get $T'_y(\bar{x}) = \lambda_y(\bar{x})\,T'_{\bar{y}}(\bar{x})$ and, owing to the linearity of T'_y, λ_y does not depend on $\bar{x}$. Hence we have $T_y(\bar{x}) = \lambda_y \bar{z} \otimes w$. For each y, we can rescale w, so that we write $T_y(\bar{x}) = \bar{z} \otimes w$. The left-hand side depends linearly on y. Hence, there is a linear mapping $Y \xrightarrow{V} W$ such that $T_y(\bar{x}) = \bar{z} \otimes Vy$. If we let $\bar{x}$ vary in X, the left-hand side depends linearly on $\bar{x}$. This entails that there is a linear mapping $X \xrightarrow{U} Z$ such that $T(x \otimes y) = Ux \otimes Vy$. This means that $T = U \otimes V$. If we exploit the canonical isomorphism $W \otimes Z \xrightarrow{\eta} Z \otimes W$, the second case can be reduced to the first, so that we have $T = \eta \circ (U \otimes V)$, where now $X \xrightarrow{U} W$ and $Y \xrightarrow{V} Z$.

In order to summarize the results of the above analysis, we introduce the following notations and definitions. If Z and W are linear spaces, and $z \in Z, w \in W$, we denote ι_w the injection

$$Z \xrightarrow{\iota_w} Z \otimes W$$

defined by the position $z \mapsto z \otimes w$ and ι_Z the injection

$$W \xrightarrow{\iota_z} Z \otimes W$$

defined by the position $w \mapsto z \otimes w$.

Definition C.2 If $X \otimes Y \xrightarrow{T} Z \otimes W$ is a linear mapping which preserves indecomposable elements, we say that T is of the first kind if there are a linear mapping $X \otimes Y \xrightarrow{A} Z$ and $w \in W$ such that $T = \iota_w \circ A$ or a linear mapping $X \otimes Y \xrightarrow{B} W$ and $z \in Z$ such that $T = \iota_z \circ B$.

Definition C.3 If $X \otimes Y \xrightarrow{T} Z \otimes W$ is a linear mapping which preserves indecomposable elements, we say that T is of the second kind if it is not of the first kind.

Theorem C.7 *Every mapping T of the second kind is the tensor product $U \otimes V$ of $X \xrightarrow{U} Z$ and $Y \xrightarrow{V} W$ or the composition $\eta \circ \left(U' \otimes V'\right)$ of the tensor product $U' \otimes V'$ of $X \xrightarrow{U'} W$ and $Y \xrightarrow{V'} Z$ with the canonical isomorphism $W \otimes Z \xrightarrow{\eta} Z \otimes W$.*

Remark C.2 In the particular case of a surjective linear mapping which preserves indecomposable elements, this mapping is necessarily of the second kind (provided that $Z \otimes W$ is not trivial). Indeed in this case ι_w (or ι_z) is strictly injective.

Suppose now that there is a linear space isomorphism T between $X \otimes Y$ and $Z \otimes W$ preserving indecomposable elements. A necessary condition for its existence is that $X \otimes Y$ and $Z \otimes W$ have the same dimension. Furthermore, on the basis of the above remark, it must be of the second kind. Consider the first case of Theorem II SP. Both U and V must be isomorphisms. Indeed, the dimension of the image of $U \otimes V$ is the product of the dimensions of the images of U and V, so that if at least one of them is not surjective, the dimension of this image is strictly lower than the dimension of $Z \otimes W$. Furthermore, if the kernel of U or V is nonzero, $U \otimes V$ too has a nontrivial kernel. Conversely, if U and V are isomorphisms, $U \otimes V$ is an isomorphism. Reasoning in the same way after having exchanged the roles of Z and W, we conclude that U' and V' are isomorphisms. Hence we obtain the following.

Theorem C.8 *If $X \otimes Y \xrightarrow{T} Z \otimes W$ is a bijective linear mapping which preserves indecomposable elements, it is the tensor product $U \otimes V$ of $X \xrightarrow{U} Z$ and $Y \xrightarrow{V} W$ where U and V are isomorphisms, or the composition $\eta \circ \left(U' \otimes V'\right)$ of the tensor product $U' \otimes V'$ of $X \xrightarrow{U'} W$ and $Y \xrightarrow{V'} Z$ with the canonical isomorphism $W \otimes Z \xrightarrow{\eta} Z \otimes W$ where U' and V' are isomorphisms.*

Appendix D
Mathematical Frameworks

D.1 Hilbert Spaces

It is useful[1] to regard the Hilbert spaces as members of the category of linear spaces over $\mathbb{C}$. We recall that the objects of this category are the linear spaces themselves and that the morphisms are the linear maps. We first observe that $\mathbb{C}$ can be regarded as a linear space over itself. The elements of $Hom\,(L, \mathbb{C})$ (i.e., the linear functionals on the linear space L) form a set L^d which has the natural structure of a linear space. We will therefore put $L^d = Hom(L, \mathbb{C})$ and equip it with this structure. We recall that L^d is called the (*algebraic*) dual of L.

If M is a linear space and $\varphi \in Hom\,(L, M)$, we can define $\varphi^d \in Hom\left(M^d, L^d\right)$ as follows. If $z \in M^d$, $z \circ \varphi \in L^d$. The corresponding mapping from M^d to L^d (for given φ) is linear and we denote it φ^d. This mapping will be called the transpose of φ. Furthermore, φ^d depends linearly on φ. Consequently, a canonical linear mapping is defined from $Hom\,(L, M)$ to $Hom\left(M^d, L^d\right)$, but for infinite-dimensional spaces it is not an isomorphism. It is easy to prove the following rule of calculation:

$$\text{if } \varphi \in Hom\,(L, M) \text{ and } \psi \in Hom\,(M, N), \quad (\psi \circ \varphi)^d = \varphi^d \circ \psi^d .$$

A Hilbert space H is a linear space over $\mathbb{C}$ equipped with a scalar product $\langle\ |\ \rangle$ which is a Banach space under the norm $\|x\| = (\langle x|x\rangle)^{1/2}$. We recall the algebraic properties of the scalar product:

$$\langle \lambda x|y\rangle = \lambda^* \langle x|y\rangle , \quad \langle x|\lambda y\rangle = \lambda \langle x|y\rangle , \quad \langle y|x\rangle = \langle x|y\rangle^* .$$

---

[1]The content of this chapter can be found in the Campanella's files *qubit1* (03/08/2011), *measurements* (26/10/2011), *stabilizer* (25/05/2014)

M. Campanella et al., *Interpretative Aspects of Quantum Mechanics*, UNIPA Springer Series,
https://doi.org/10.1007/978-3-030-44207-1

Furthermore, the scalar product is additive with respect to each argument. The norm induces in H a topology. If $\mathbb{C}$ is equipped with its standard topology, we can select in H^d the subset H^* of the continuous functionals of H, which is a linear space. This space will be called the *topological dual of* H. In what follows, unless differently specified, the dual of a Hilbert space must be understood as the topological dual.

D.1.1 Linear Mappings Between Hilbert Spaces

Two basic facts are known from functional analysis. The first is that a linear mapping from a Hilbert space H to a Hilbert space K is continuous (with respect to their norm-induced topologies) if and only if it is bounded. In particular, a linear functional on a Hilbert space is continuous if and only if it is bounded. The second is the Riesz representation theorem which states the following.

Riesz Representation Theorem: *In a Hilbert space H, for each bounded functional φ on H there is a unique $x \in H$ such that $\varphi = \langle x| \ \rangle$. Furthermore, for each $x \in H$ the linear functional $\langle x| \ \rangle$ is bounded.*

An immediate consequence of this theorem is that for a Hilbert space H there is a *canonical bijection* between H and H^*. The element of H^* which corresponds to $x \in H$ will be denoted with x^* and is called the *conjugate* of x. This canonical bijection is additive, and furthermore $(\lambda x)^* = \lambda^* x^*$. When a mapping enjoys such properties, we say that it is *antilinear*. In our case, the mapping is bijective and consequently we say that it is an *anti-isomorphism*.

A mapping f from a normed space X to a normed space Y is called an *isometry* if it preserves the norm, that is, if $\|f(x)\| = f(\|x\|), \forall x \in X$.

We now prove the following theorem.

Theorem D.1 *There is on H^* a unique structure of Hilbert space such that * is an antilinear isometry. The scalar product for this structure is defined by*

$$\langle x^* | y^* \rangle_{H^*} = \langle y | x \rangle .$$

Proof We will base the proof on the following lemma.

Lemma D.1 *If X is a Hilbert space and $\langle \ | \ \rangle'$ is a scalar product such that the identity map ι is an isometry between the normed spaces $(X, \| \ \|)$ and $(X, \| \ \|')$, then $\langle \ | \ \rangle = \langle \ | \ \rangle'$.*

Proof of the Lemma: Owing to the assumption that ι is an isometry, we must have $\langle x|x\rangle = \langle x|x\rangle'$. Using the properties of the scalar product, it is easy to prove the following identity:

$$\langle x|y\rangle = -\tfrac{1}{4}\left(\langle x-y|x-y\rangle - \langle x+y|x+y\rangle + \langle x-iy|x-iy\rangle - \langle x+iy|x+iy\rangle\right)$$

so that $\langle\ |\ \rangle$ is uniquely defined by $\|\ \|$. Therefore $\|\ \| = \|\ \|' \Rightarrow \langle\ |\ \rangle = \langle\ |\ \rangle'$. □

Proof of the theorem: We define $\langle x^*|y^*\rangle_{H^*} = \langle y|x\rangle_H$. We have:

$$\left\langle x^*|\lambda\left(y^*\right)\right\rangle_{H^*} = \left\langle x^*|\left(\lambda^* y\right)^*\right\rangle_{H^*} = \left\langle \lambda^* y|x\right\rangle_H = \lambda\langle y|x\rangle_H = \lambda\left\langle x^*|y^*\right\rangle_{H^*}.$$

Thus $\langle\ |\ \rangle_{H^*}$ is linear with respect to the right argument. In a similar way, it can be shown that it is antilinear with respect to the left argument. The other axioms of a scalar product are obviously satisfied. Furthermore, we have

$$\left\|x^*\right\|_{H^*} = \left\langle x^*|x^*\right\rangle_{H^*}^{1/2} = \langle x|x\rangle_H^{1/2} = \|x\|_H$$

so that * is indeed an isometry. Finally, if $\|\ \|_{H^*}'$ is the norm induced by another scalar product $\langle\ |\ \rangle_{H^*}'$, we must have $\|x^*\|_{H^*}' = \|x\|_H = \|x^*\|_{H^*}$ so that, using the lemma, $\langle\ |\ \rangle_{H^*}'$ must be equal to $\langle\ |\ \rangle_{H^*}$. □

In what follows, it will be understood that the dual of a Hilbert space is a Hilbert space with the scalar product introduced above.

A twofold application of * to H defines a canonical isomorphism between H and H^{**}. Passing to the quotient category modulo all these isomorphisms, we identify each Hilbert space with its double dual, so that $(x^*)^* = x$. We have

$$x^{**}\left(z^*\right) = \left\langle x^*|z^*\right\rangle_{H^*} = \langle z|x\rangle\,.$$

Therefore, the vector x is identified with a linear functional over H^*, and we write $x\left(z^*\right) = \langle z|x\rangle$.

Let H and K be Hilbert spaces. If $\varphi \in Hom\,(H, K)$, then $\varphi^d \in Hom\left(K^d, H^d\right)$ is uniquely defined. It is not true in general that the restriction of φ^d to K^* has its image in H^* but, if it does, a linear mapping from K^* to H^* is defined. It can be shown that this happens if and only if φ is bounded. Henceforth, we suppose, unless differently specified, that all morphisms and antimorphisms between Hilbert spaces are bounded.

With an abuse of notation, we continue to denote $Hom\,(H, K)$ the space of bounded linear maps from the Hilbert space H to the Hilbert space K. So, in the category of Hilbert spaces, the morphisms will be taken as the bounded linear maps. With these specifications in mind, the restriction of φ^d to K^* belongs to $Hom\,(K^*, H^*)$. This restriction will be denoted $\widehat{\varphi}$ and we continue to call it the *transpose* of φ. The association of $\widehat{\varphi}$ with each φ defines a canonical linear map from $Hom\,(H, K)$ to $Hom\,(K^*, H^*)$.

If we define

$$\varphi^* = (*) \circ \widehat{\varphi} \circ (*)\,,$$

we obtain an element of $Hom\,(K, H)$. The association of φ^* with each φ defines a canonical antilinear map from $Hom\,(H, K)$ to $Hom\,(K, H)$. We call φ^* the *Hermitean conjugate* (or also the *adjoint*) of φ. We easily obtain the calculation rule

$$(\varphi_1\varphi_2)^* = \varphi_2{}^*\varphi_1{}^*.$$

We now introduce the bra-ket notation. If $f \in H^*$, there is a unique $x \in H$ such that $f(y) = \langle x|y\rangle$, $\forall y \in H$, so that it is natural to put $f(\ ) = \langle x|\ \rangle$ or else $f = \langle x|$. In the following, this will be the notation for an element of H^*. Comparing the previous notation with the actual one, we have $x^* = \langle x|$. Furthermore, the identification of H with H^{**} leads to the equation $x(z^*) = \langle z|x\rangle$, which can be written $x(\langle z|) = \langle z|x\rangle$ in the new notation for the elements of H^*. It is then natural to put $x = |x\rangle$.

If H is a Hilbert space, an element of it will be henceforth called a ket and denoted $|x\rangle$, while an element of H^* will be called a bra and denoted $\langle z|$. In this notation, $|x\rangle^* = \langle x|$. From $|x\rangle^{**} = |x\rangle$ we derive $\langle x|^* = |x\rangle$. We have $\langle x|y\rangle = \langle |x\rangle \,|\, |y\rangle\rangle$.

Of course, the angular brackets are referred to the scalar product in H, so that we should write more precisely ${}_H\langle x|$ and $|y\rangle_H$. Such a level of specification is sometimes important. For example, $|x\rangle_H$ is a vector of H and ${}_H\langle x|$ the corresponding bra, regarded as a linear functional of H, but the same element can be regarded as a vector of H^* and it must consistently be written as a ket of H^*, so that we must write ${}_H\langle x| = |x\rangle_{H^*}$. But $\langle y^*|x^*\rangle_{H^*} = \langle x|y\rangle_H$, and $x^* = |x\rangle_{H^*}$, $y^* = |y\rangle_{H^*}$, so that we obtain

$${}_{H^*}\langle y|x\rangle_{H^*} = {}_H\langle x|y\rangle_H.$$

The anti-isomorphism $|x\rangle_H \mapsto {}_H\langle x|$ can be rewritten as $|x\rangle_H \mapsto |x\rangle_{H^*}$, so that we have $|x\rangle_H^* = |x\rangle_{H^*}$. A useful mnemonic for these rules is that

$$|\diamond\rangle_H^* = {}_H\langle\diamond| \quad \text{and} \quad |\diamond\rangle_H^* = |\diamond\rangle_{H^*}.$$

In this case, the complete notation is essential. However, we will use the simplified notation when no ambiguity can arise.

Let φ be an element of $Hom\,(H, K)$; the element $\widehat{\varphi}$ belongs to $Hom\,(K^*, H^*)$; if $\langle y| \in K^*$ the composition of maps $\langle y| \circ \varphi$ is a functional of H that is nothing but $\widehat{\varphi}(\langle y|)$. If we omit the symbol for the composition of maps, we can describe the action of $\widehat{\varphi}$ as follows:

$$K^* \xrightarrow{\widehat{\varphi}} H^* \;:\; \langle y| \mapsto \langle y|\,\varphi.$$

We have therefore the rule $\widehat{\varphi}(\langle y|) = \langle y|\,\varphi$.

Using the above notation, *the same symbol corresponds to two different objects: when it operates on the left, it represents some morphism, while when it operates on the right, it represents its transpose. In order to avoid ambiguities, we use the convention that the specification of the source and of the target refers to the action on the left. Namely, if we write $\varphi \in Hom\,(H, K)$ we intend that when φ acts on the left on an element of H, it produces an element of K, while its action on the right* *on an element of K^* produces an element of H^* according to the transpose mapping.*

Let us now consider $\varphi^* = (*)\,\widehat{\varphi}\,(*)$. It operates as follows:

$$|y\rangle \mapsto \langle y| \mapsto \langle y|\,\varphi \mapsto \varphi^*\,|y\rangle\,.$$

If we put $|z\rangle = \varphi^*\,|y\rangle$, then $\langle z| = \langle y|\,\varphi$. It is easy to see that $\varphi^{**} = \varphi$, so that we also have

$$|z\rangle = \varphi\,|y\rangle \;\Rightarrow\; \langle z| = \langle y|\,\varphi^*.$$

We emphasize that in the above equation φ^* operates on the right, so its action is that of the transpose morphism.

Summarizing, if $\varphi \in Hom\,(H, K)$, $\varphi^* \in Hom(K, H)$, but the action of φ^* on the left leads from K to H and the action of φ^* on the right leads from H^* to K^*.

The complex field, as already observed, can be regarded as a linear space over $\mathbb{C}$. This means that $\mathbb{C}$ is regarded as a $\mathbb{C}$-module, i.e., as the additive group $\mathbb{C}$ whose selected endomorphisms are the homotheties $c \mapsto \lambda c$, $c \in \mathbb{C}$, $\lambda \in \mathbb{C}$. This linear space will be denoted $L_{\mathbb{C}}$. We can introduce in it a natural scalar product $\langle c|c'\rangle = c^*c'$, and it becomes of course a Hilbert space under this product. This space will be denoted $H_{\mathbb{C}}$. A linear functional on $H_{\mathbb{C}}$ is uniquely expressed through the position $c \to a^*c$ for some $a \in H_{\mathbb{C}}$. This position defines the canonical anti-isomorphism between $H_{\mathbb{C}}$ and its dual $H_{\mathbb{C}}{}^*$. As the underlying set of $H_{\mathbb{C}}$ is $\mathbb{C}$, we can consider the identity map of $H_{\mathbb{C}}$ as a map from $H_{\mathbb{C}}$ to $\mathbb{C}$, which is obviously linear, so that this map is an element of $H_{\mathbb{C}}{}^*$. The corresponding ket is nothing but the unit element of the field $\mathbb{C}$, considered as an element of $H_{\mathbb{C}}$. The space $H_{\mathbb{C}}$ is obviously one-dimensional. The unit element of $\mathbb{C}$ is a selected orthonormal basis of $H_{\mathbb{C}}$. Every $x \in H_{\mathbb{C}}$ can be uniquely represented as $x = c1$, $c \in \mathbb{C}$. The identity map $\mathbb{C} \overset{id}{\to} H_{\mathbb{C}}$ is an isomorphism of additive groups. If we define a product in $H_{\mathbb{C}}$ according to the rule $c1 \cdot c'1 = cc'1$, $H_{\mathbb{C}}$ becomes a field and id an isomorphism of fields.

More generally, let us consider a one-dimensional Hilbert space. By restriction of scalars to $\mathbb{R}$ we obtain a linear space L over $\mathbb{R}$. If w is a nonzero vector, every $z \in H$ can be written in a unique way as $z = (x + iy)\,w = xw + y\,(iw)$, so that L is two-dimensional. The expression $\mathrm{Re}\,\langle z|z'\rangle$ defines a scalar product in L. Therefore, L is equipped with a structure of real Hilbert space. The norms of the two spaces are the same. The expression $\mathrm{Im}\,\langle z|z'\rangle$ defines a bilinear skew-symmetric form. The angle ϑ between the vectors z and z' (with $0 \le \vartheta \le \pi$) is defined by

$$\cos\vartheta = \frac{\mathrm{Re}\langle z|z'\rangle}{\langle z|z\rangle^{1/2}\,\langle z'|z'\rangle^{1/2}}$$

It can be regarded as an oriented Euclidean plane with a selected circle. $\mathbb{C}_H$ is obtained by selecting a point on this circle.

We may also consider an oriented Euclidean plane with a selected point. The orbit of each point under its structure group is a circle around the selected point. It becomes a Hilbert one-dimensional space as long as a specific circle around the selected point is chosen.

If H is a Hilbert space, its dual can be regarded as $Hom(H, \mathbb{C}_H)$. In this way, each bra is the adjoint of a well-defined $\varphi \in Hom\,(\mathbb{C}_H, H)$. We can then write $\langle x| = \varphi^*$. Let us denote $|1\rangle$ the selected vector in $\mathbb{C}_H$. The corresponding bra is the map $\mathbb{C}_H \xrightarrow{\iota} \mathbb{C}$ which operates as identity on the underlying set. The bra corresponding to $\varphi\,|1\rangle$ is $\iota\varphi^* = \langle x|$. We conclude that $|x\rangle = \varphi\,|1\rangle$.

We can therefore associate in a unique way an element of $Hom\,(\mathbb{C}_H, H)$ with each ket of H and its adjoint with the corresponding bra. In this way, the elements of a Hilbert space (kets) can be represented by morphisms starting from $\mathbb{C}_H$, while their corresponding bras are represented by morphisms arriving at $\mathbb{C}_H$. The canonical anti-isomorphism corresponds to the passage to the adjoint morphism. If $|y\rangle = \psi\,|1\rangle$ and $|x\rangle = \varphi\,|1\rangle$, $\langle x|y\rangle = \varphi^*\psi\,|1\rangle$.

If $H = \mathbb{C}_H$, each element of $\mathbb{C}_H$ is represented by an element of $Hom\,(\mathbb{C}_H, \mathbb{C}_H)$, i.e., as an element of $\mathbb{C}$ regarded as a multiplier. We can write $|c\rangle = c\,|1\rangle$. In this representation, $\mathbb{C}_H$ is identified with $\mathbb{C}_H^*$. Therefore, besides the canonical anti-isomorphism between $\mathbb{C}_H$ and $\mathbb{C}_H^*$ (which in this representation corresponds to complex conjugation), there is a canonical isomorphism which in this representation corresponds to identity.

Let $\varphi \in Hom\,(H, \mathbb{C}_H)$. Its transpose is an element of $Hom\left(\mathbb{C}_H^*, H^*\right)$, that is, an element of $Hom\,(\mathbb{C}_H, H^*)$. According to the rules involving the transpose,

$$\widehat{\varphi}\,(\langle c|) = \langle c|\,\varphi.$$

D.1.2 Tensor Products. Universal Property

Let us now come to tensor products between linear spaces. If L, M, and N are linear spaces, a mapping from $L \times M$ to N is called bilinear if it is linear with respect to each of its arguments. It is well known that, for given L and M, there is a linear space $L \otimes M$ together with a bilinear mapping $\otimes$ from $L \times M$ to $L \otimes M$ such that every bilinear mapping ψ from $L \times M$ to any N can be uniquely factorized as $\psi = \varphi \circ \otimes$ where φ is a linear mapping from $L \otimes M$ to N. The space $L \otimes M$ is determined up to isomorphisms and is called the (algebraic) tensor product of L and M.

A specific construction of $L \otimes M$ is well known. Starting from the set $L \times M$ we build the free linear space over $L \times M$ (i.e., the linear space of all formal finite linear combinations of elements of $L \times M$ with coefficients in $\mathbb{C}$), then we pass to the quotient linear space modulo the subspace generated by all elements of the form

$$(l + l', m) - (l, m) - (l', m),$$

$$(l + l', m) - (l, m) - (l', m),$$

$$(\lambda l, m) - \lambda(l, m),$$

$$(\lambda l, m) - \lambda(l, m),$$

with $m, m' \in M, n, n' \in N, \lambda \in \mathbb{C}$. This particular realization of $L \otimes M$ will be called the *standard tensor product*. Unless differently specified, by (algebraic) tensor product of two linear spaces we will understand their standard tensor product.

We observe that there is a canonical isomorphism between $L \otimes M$ and $M \otimes L$. In fact $\otimes$ for the second tensor product can be regarded a bilinear map from $L \times M$ to $M \otimes L$, so that there is a unique associated linear map from $L \otimes M$ to $M \otimes L$. By repeating this argumentation with the roles of L and M interchanged, we conclude that this map is an isomorphism. The property of the tensor product of uniquely associating a linear map to an arbitrary bilinear map of fixed domain is known as *the universal property*. The elements of the tensor product admitting a representation $l \otimes m$ are called *indecomposable*.

We now define the tensor product of two Hilbert spaces. Let H, K be Hilbert spaces and consider their tensor product $H \otimes K$ as linear spaces. For indecomposable elements $h \otimes k$ and $h' \otimes k'$, we define their scalar product as

$$\langle h \otimes k | h' \otimes k' \rangle = \langle h | h' \rangle \langle k | k' \rangle .$$

As any element of $H \otimes K$ can be expressed as a finite sum $\sum_i h_i \otimes k_i$, we can extend the definition to the whole space by bi-additivity. The completion of the pre-Hilbertian space so obtained is by definition the (topological) tensor product of the given Hilbert spaces. It is obvious that the canonical isomorphism between $H \otimes K$ and $K \otimes H$ as algebraic tensor products is inherited by their topological counterparts.

A natural question now arises: in what measure the universal property of tensor products for linear spaces can be extended to Hilbert spaces?

Let H, K be Hilbert spaces. It is natural to consider only continuous bilinear maps. If $H \times K \xrightarrow{\Upsilon} L$ is a bilinear map toward a Hilbert space L, we say that Υ is bounded if there is $M > 0$ such that

$$\| \Upsilon (|x\rangle , |y\rangle) \| < M \|x\| \|y\| \quad \forall |x\rangle \in H, |y\rangle \in K.$$

The natural topology of $H \times K$ as a Cartesian product is the product topology, which is induced by the norm $\|x\| + \|y\|$, while its topology as direct sum $H \oplus K$ is induced by the Hilbert space norm $\| (|x\rangle , |y\rangle) \| = \left(\|x\|^2 + \|y\|^2 \right)^{1/2}$.

Suppose that a sequence $(|x_n\rangle, |y_n\rangle) \to (|x\rangle, |y\rangle)$ in the product topology. This means that for any $\varepsilon > 0$ there is a natural N such that

$$\| \, |x_n\rangle - |x\rangle \| + \| \, |y_n\rangle - |y\rangle \| < \varepsilon \qquad \text{for } n > N.$$

But

$$(\langle x_n - x | x_n - x\rangle + \langle y_n - y | y_n - y\rangle)^{1/2} \leq \|x_n - x\| + \|y_n - y\| ,$$

so that $(|x_n\rangle, |y_n\rangle) \to (|x\rangle, |y\rangle)$, in the Hilbert space topology. Vice versa, suppose that $(|x_n\rangle, |y_n\rangle) \to (|x\rangle, |y\rangle)$ in the Hilbert space topology. As

$$\|x_n - x\| + \|y_n - y\| \leq 2 \max (\|x_n - x\| , \|y_n - y\|) \leq 2\left(\|x_n - x\|^2 + \|y_n - y\|^2\right)^{1/2},$$

$(|x_n\rangle, |y_n\rangle) \to (|x\rangle, |y\rangle)$ in the product topology. Therefore, the two topologies coincide.

We now show that, if Υ is bounded, it is continuous. Indeed, we have

$$\|\Upsilon(x_n, y_n) - \Upsilon(x, y)\| = \|\Upsilon(x_n - x, y_n) + \Upsilon(x, y_n) - \Upsilon(x, y)\| .$$

But $\Upsilon(x_n - x, y_n) = \Upsilon(x_n - x, y_n - y) + \Upsilon(x_n - x, y)$. Therefore, we have

$$\|\Upsilon(x_n, y_n) - \Upsilon(x, y)\| = \|\Upsilon(x_n - x, y) + \Upsilon(x, y_n - y) + \Upsilon(x_n - x, y_n - y)\| ,$$

whence

$$\|\Upsilon(x_n, y_n) - \Upsilon(x, y)\| \leq M \|x_n - x\| \|y\| + M \|x\| \|y_n - y\| + M \|x_n - x\| \|y_n - y\| ,$$

from which we see that Υ is continuous. Conversely, suppose that Υ is continuous. Let us consider its restriction to $S \times S'$, where S and S' are the unit spheres of H and K. As they are compact, $\|\Upsilon\|$ has a maximum; therefore, the bilinearity implies the boundedness of Υ.

We now afford the question of the universality of the tensor product of two Hilbert spaces. Let $H, K, and L$ be Hilbert spaces and $H \times K \xrightarrow{\Upsilon} L$ a bounded bilinear map. Let us for the moment $H \otimes_A K$ denote the algebraic tensor product. If there is a continuous linear mapping $H \otimes K \xrightarrow{\varphi} L$ such that $\Upsilon = \varphi \circ \otimes$, its restriction $\varphi_{|A}$ to $H \otimes_A K$ is uniquely defined by the universality property of algebraic tensor products, and it satisfies $\Upsilon = \varphi_{|A} \circ \otimes$. As $H \otimes_A K$ is dense in $H \otimes K$, each $w \in H \otimes K$ is the limit of some sequence $w_n \in H \otimes_A K$. The continuity of φ implies that $\varphi(w) = \lim_{n \to \infty} \varphi_{|A}(w_n)$. Therefore, if φ exists, it is unique. Let us show now that φ exists. We first get a unique φ_A from Υ by virtue of the universality property of the algebraic tensor product; the domain of φ_A is the algebraic tensor product. The boundedness of Υ implies that of φ_A; therefore, the latter can be extended to $H \otimes K$ which is the adherence of $H \otimes_A K$. In this way, we have proved the following theorem.

Theorem D.2 *The tensor product of two Hilbert spaces satisfies the universality property, that is, for every bounded bilinear mapping* $H \times K \xrightarrow{\Upsilon} L$ *there is a unique bounded linear mapping* $H \otimes K \xrightarrow{\varphi} L$ *such that* $\Upsilon = \varphi \circ \otimes$.

D.2 Canonical Factorization

Another important item which will be useful in the following is the canonical factorization of a morphism. We recall that, if L and M are linear spaces and $\varphi \in Hom\,(L, M)$, the following unique factorization holds:

$$L \xrightarrow{\pi} L/\ker\varphi \xrightarrow{\vartheta} \mathrm{Im}\,\varphi \xrightarrow{\iota} M,$$

where π is the canonical projection over the quotient space, ϑ is an isomorphism, and ι the insertion morphism.

If $L = H$ and $M = K$ are Hilbert spaces, we want to introduce scalar products in $L/\ker\varphi$ and in $\mathrm{Im}\,\varphi$ in such a way that they become Hilbert spaces and all the morphisms are continuous. To this purpose the following considerations will be useful.

If Y is a Hilbert space and $X \xrightarrow{\iota} Y$ is an injective linear mapping whose image is a closed subspace of Y, there is a unique scalar product in X such that ι is an isometry with respect to the induced norm in X. This unique scalar product is defined by $\langle x_1 | x_2 \rangle = \langle \iota\,(x_1)\, |\, \iota\,(x_2) \rangle$. Furthermore X is complete, so that it becomes a Hilbert space.

Once a structure of Hilbert space has been defined in X, we can rewrite the isometry condition as $\langle x_1 | x_2 \rangle = \langle x_1 |\, \iota^* \iota\, | x_2 \rangle$, so that $\iota^* \iota = Id_X$. The operator $\iota\iota^*$ acting on Y is self-adjoint; furthermore, $\iota\iota^* \iota\iota^* = \iota\iota^*$, so that it is a projector of Y. The eigenspace of $\iota\iota^*$ associated with the eigenvalue 0 is characterized by the equation $\iota\iota^* \,|y\rangle = 0$. Multiplying on the left by $\langle y|$ we get $\langle y|\, \iota\iota^* \,|y\rangle = 0$, which implies $\langle y|\, \iota = 0$. Vice versa, $\langle y|\, \iota = 0$ implies $\iota^* \,|y\rangle = 0$ and then $\iota\iota^* \,|y\rangle = 0$. We therefore have

$$\iota\iota^* \,|y\rangle = 0 \;\Leftrightarrow\; \langle y|\, \iota = 0 \Leftrightarrow \forall\, |z\rangle \in Y\; \langle y|\, \iota\, |z\rangle = 0.$$

This means that the null space of $\iota\iota^*$ is nothing but $Im(\iota)_\perp$. Then the other eigenspace of the projector is $Im(\iota)$, so that $\iota\iota^*$ projects on $Im(\iota)$. The following lemma will prove useful.

Lemma D.2 *If* $L \xrightarrow{\pi} M$ *is a surjective linear mapping, its transpose* $M^d \xrightarrow{\pi^d} L^d$ *is injective.*

Proof If $z \in M^d$ and $y \in L$, then $\pi^d(z)(y) = z(\pi(y))$. As π is surjective, $\pi^d(z) = 0 \Rightarrow z = 0$ so that π^d is injective.

If L and M are Hilbert spaces, we have the following.

Lemma D.3 *If $L \xrightarrow{p} M$ is surjective, $M \xrightarrow{p^*} L$ is injective, and if $L \xrightarrow{\iota} M$ is injective, $M \xrightarrow{\iota^*} L$ is surjective.*

Proof If $p^*|x\rangle = 0$, then $\langle y| p^* |x\rangle = 0 \forall |y\rangle \in L$, that is, $\langle x| p |y\rangle = 0 \forall |y\rangle \in L$. But $p|y\rangle$ is an arbitrary vector of M, so that $|x\rangle = 0$. We have $\iota^*\iota = Id_L$, so that $|x\rangle = \iota^*\iota|x\rangle \,\forall |x\rangle \in L$ and then ι^* is surjective.

Let us suppose now that L is a Hilbert space and $L^* \subseteq L^d$ its topological dual. We also suppose that L is a Hilbert space, M is a linear space, $L \xrightarrow{\pi} M$ is surjective, and $\ker \pi$ is a closed subspace of L. Define

$$\widetilde{M} = \left(\pi^d\right)^{-1}(L^*).$$

The restriction η of π^d to $\widetilde{M}$ is an injective linear mapping $\widetilde{M} \xrightarrow{\eta} L^*$. As a consequence of the assumption that $\ker \pi$ is closed, it can be shown that $\mathrm{Im}\,\eta$ is a closed subspace of L^*. Therefore, a unique structure of Hilbert space exists in $\widetilde{M}$ such that η is an isometry. The corresponding scalar product is $\langle u|v\rangle_{\widetilde{M}} = \langle \eta(u)|\eta(v)\rangle_{L^*}$ $\forall u, v \in \widetilde{M}$. The adjoint η^* of η is surjective and leads from L^* to $\widetilde{M}$. But $\widetilde{M}$ is contained in M^d, so that each $x \in \widetilde{M}$ is a functional of M. Therefore, a bilinear mapping $\widetilde{M} \times M \xrightarrow{\Upsilon} \mathbb{C}$ is defined as $\Upsilon(x, y) = x(y) \forall x \in \widetilde{M}, y \in M$. For each $y \in M$, a linear functional of $\widetilde{M}$ arises, i.e., $\Upsilon(\ , y)$. It can be shown that this functional is bounded, so that a linear map from M to $\widetilde{M}^*$ is defined.

Vice versa, if $\langle u| \in \widetilde{M}^*$, a bounded linear functional $\widetilde{u}$ of $\widetilde{M}$ is defined. But $\eta(\widetilde{u}) \in L^*$, i.e., it is a bra $\langle \widetilde{u}|$, so that $|\widetilde{u}\rangle \in L$ and $\pi(|\widetilde{u}\rangle) \in M$. There is then a bijection between $\widetilde{M}^*$ and M. This bijection is manifestly antilinear. If we compose it with the standard anti-isomorphism between $\widetilde{M}$ and $\widetilde{M}^*$, we get a linear space isomorphism between M and $\widetilde{M}$.

The construction outlined above shows that this isomorphism is canonical. By means of this isomorphism we can carry the Hilbert space structure from $\widetilde{M}$ to M. We have just proved the following.

Theorem D.3 *If $L \xrightarrow{\pi} M$ is a surjective linear mapping, L is a Hilbert space and* $\ker \pi$ *is a closed subspace of L, there is a unique Hilbert space structure on M such that the restriction of π^d to $\left(\pi^d\right)^{-1}(L^*)$ is an isometry.*

Let us consider the adjoint morphism π^*. We recall that $\pi^* = (*)\,\widehat{\pi}\,(*)$, where $\widehat{\pi}$ is the restriction of π^d to M^*. Let $|y\rangle$ be a ket of M. The passage to $\langle y|$ is isometric. This bra is a bounded functional of M, that is, an element of $\widetilde{M}$; we recognize that $\widehat{\pi}$ in our case is nothing but η which is an isometry. The final application of $(*)$ does not change the norm. We conclude that π^* is an isometry. We can therefore write that $\langle y| \pi\pi^* |y\rangle = \langle y|y\rangle \, \forall y \in M$. Observing that $\pi\pi^*$ is self-adjoint, we deduce that $\pi\pi^* = Id_M$.

Furthermore, also $\pi^*\pi$ is self-adjoint, and $\pi^*\pi\pi^*\pi = \pi^*\pi$, so that $\pi^*\pi$ is a projector operating in L. The eigenspace of this projector associated with the eigenvalue zero is $\ker \pi$, while the eigenspace associated with 1 is its orthogonal complement, so that $\pi^*\pi$ projects on $\ker \pi_\perp$. The whole L is the direct sum of these subspaces. If υ is the insertion of $\ker \varphi_\perp$ in L, $\pi\upsilon$ is a canonical isometry between $\ker \varphi_\perp$ and M.

We can summarize all the previous considerations in the following.

Theorem D.4 *If $X \xrightarrow{\iota} Y$ is an injective linear mapping, Y is a Hilbert space and* $\operatorname{Im} \iota$ *is closed in Y, there is in X a unique structure of Hilbert space such that ι is an isometry. With respect to this structure $\iota^*\iota = Id_X$ and $\iota\iota^*$ is a projector acting on Y and projecting on* $\operatorname{Im} \iota$.

If $L \xrightarrow{\pi} M$ is a surjective linear map, L is a Hilbert space and $\ker \pi$ *is closed in L, there is in M a unique structure of Hilbert space such that π^* is an isometry. We have $\pi\pi^* = Id_M$ and $\pi^*\pi$ is a projector acting on L and projecting on* $\ker \pi_\perp$. *If υ is the insertion of* $\ker \pi_\perp$ *in L, $\pi\upsilon$ is a canonical isometry between* $\ker \pi_\perp$ *and M.*

Returning to the canonical factorization, and remembering that π is surjective and ι is injective, we will equip $L/\ker \varphi$ and $\operatorname{Im} \varphi$ with their natural structures of Hilbert spaces, so that all the morphisms in the factorization will be here on regarded as Hilbert space morphisms. In particular, their adjoints are defined and they will be very useful in the following.

We will now prove the following theorem which establishes a connection between tensor products and morphisms.

Theorem D.5 *There is a canonical linear space injective morphism from the tensor product $H \otimes K$ to $Hom\,(H^*, K)$. There is a unique structure of Hilbert space on its image such that this injection is an isometry.*

Proof For the universal property of the tensor product, there is a canonical linear space isomorphism between the space of bounded linear functionals of $H \otimes K$ and the space of bounded bilinear maps of $H \times K$ in $\mathbb{C}$. But each linear bounded functional of $H \otimes K$ is a bra whose associated ket depends antilinearly on the functional. In this way, we get a canonical anti-isomorphism between $H \otimes K$ and the space of bounded bilinear maps of $H \times K$ in $\mathbb{C}$ and hence an isomorphism between $H \otimes K$ and the linear space of the conjugate maps. These maps are bi-antilinear. For each map, the assignment of any vector $|x\rangle$ of H defines a bounded antilinear functional of K, that is, a ket of K. This functional (i.e., the ket) depends antilinearly on $|x\rangle$ and linearly on the corresponding bra $\langle x|$, and it can be shown that this dependence is continuous, so that we get an element of $Hom\,(H^*, K)$. If we multiply an element of $H \otimes K$ by a scalar, the corresponding conjugate map is multiplied by the same scalar, so that, for given $\langle x|$, the ket is multiplied again by the same scalar. We conclude that the mapping between $H \otimes K$ and $Hom\,(H^*, K)$ is a linear space morphism, which is injective. The unique structure of Hilbert space of the image is obtained trivially by transferring the scalar product. In the canonical injection, the ket $|h\rangle \otimes |k\rangle$ goes in the morphism

$$H^* \xrightarrow{|h\rangle|k\rangle} K : \langle x| \mapsto \langle x|h\rangle\, |k\rangle\,,$$

so that $\big(|h\rangle\,|k\rangle \mid |h'\rangle\,|k'\rangle\big) = \langle h|h'\rangle\langle k|k'\rangle$. □

Remark D.1 The image of the canonical injection is the adherence in $Hom\,(H^*, K)$ of the subspace of $Hom\,(H^*, K)$ whose elements are of finite rank. This adherence is itself a subspace of $Hom\,(H^*, K)$ and it will be called the *space of Hilbert–Schmidt mappings*. We will denote this space $HS\,(H^*, K)$. We observe that if the Hilbert spaces involved are finite-dimensional, $HS\,(H^*, K) = Hom\,(H^*, K)$.

D.3 Hilbert–Schmidt Spaces

Let φ be an element of $HS\,(H^*, K)$. Its adjoint is an element of $HS\,(H^*, K)$ and the transpose of the latter is an element of $HS\,(H^*, K)$. The composition of these two operations is an anti-isomorphism. Furthermore, there is a canonical isomorphism between $H \otimes K$ and $K \otimes H$. From the above considerations, the following theorem is derived.

Theorem D.6 *There are canonical isomorphisms among* $H \otimes K$, $K \otimes H$, $HS\,(H^*, K)$ *and among* $(H \otimes K)^*$, $(K \otimes H)^*$, $HS\,(H, K^*)$. *There are canonical anti-isomorphisms between each element of the first set and each element of the second.*

As the above statements hold for arbitrary H and K, replacing them with H^* and K^*, respectively, we get a canonical isomorphism between $H^* \otimes K^*$ and $(H \otimes K)^*$.

It is convenient to introduce an efficient notation for morphisms based on bra-ket symbolism.

Let us first consider a particular kind of morphism $\varphi \in HS\,(H, K)$ whose image is a one-dimensional subspace of K. Let $|k\rangle$ be a generator of this subspace. If $|x\rangle \in H$, $\varphi\,|x\rangle$ is proportional to $|k\rangle$ according to some coefficient which depends linearly and continuously on $|x\rangle$. There is therefore a bra $\langle h| \in H^*$ such that $\varphi\,|x\rangle = |k\rangle\,\langle h|x\rangle$. We can then introduce the notation $\varphi = |k\rangle\,\langle h|$. The ket $|k\rangle$ is defined up to a coefficient but, if we multiply $|k\rangle$ by a factor, we must divide $\langle h|$ by the same factor in order to preserve φ. A morphism of this type will be henceforth called an *indecomposable morphism*.

We know that the action of φ on the right defines the transpose. It is easy to check that the above notation for φ is consistent, i.e., that the position $\langle x| \mapsto \langle x|k\rangle\,\langle h|$ defines the transpose.

We have $\langle x|\,\varphi^* = \langle x|h\rangle\,\langle k|$, so that $(|k\rangle\,\langle h|)^* = |h\rangle\,\langle k|$.

As a consequence of the isomorphism between tensor products and spaces of morphisms, an arbitrary φ can always be expanded as a (perhaps infinite) sum $\varphi = \sum_{\alpha} |k_\alpha\rangle \langle h_\alpha|$ (in a highly no unique way). Therefore, all the rules of calculation to follow will be formulated for indecomposable morphisms. When required, they will be uniquely extended by additivity.

The spaces H, K are arbitrary Hilbert spaces. Therefore, each of them can be replaced with its dual.

Let us consider $\varphi \in HS(H^*, K)$. If $\varphi = |k\rangle_K\ {}_{H^*}\langle h|$, then

$$\varphi|x\rangle_{H^*} = |k\rangle_K\ {}_{H^*}\langle h|x\rangle_{H^*} = \ {}_H\langle x|h\rangle_H|k\rangle_K.$$

Hence, considering x as a bra of H instead than a ket of H^* and putting $\varphi = |h\rangle\,|k\rangle$, we write $\langle x|\,\varphi$ instead of $\varphi|x\rangle_{H^*}$. In this way, φ operates on the right.

We further have ${}_K\langle x|k\rangle_K\ {}_{H^*}\langle h| = \ {}_K\langle x|k\rangle_K|h\rangle_H$. Therefore, considering x as a bra of K and the result as a ket of H, the transpose of φ operates on the right and is given by $|k\rangle\,|h\rangle$. The adjoint of φ is $|h\rangle_{H^*}\ {}_K\langle k|$, i.e., $\langle h|\,\langle k|$. The transpose of the adjoint operates as ${}_{H^*}\langle x|h\rangle_{H^*}\ {}_K\langle k| = \langle k|\,\langle h|x\rangle$; hence, the transpose of $\langle h|\,\langle k|$ is $\langle k|\,\langle h|$.

Summarizing, we can give a meaning to any expression consisting of a sum of terms of the three kinds $|\ \rangle\,|\ \rangle$, $|\ \rangle\langle\ |$, $\langle\ |\langle\ |$; let us call K the Hilbert space of the left variable and H the Hilbert space of the right variable. A sum of terms $|k\rangle\,|h\rangle$ is interpreted as an element of $HS(K^*, H)$ acting on K^* on the right; a sum of terms $|k\rangle\,\langle h|$ is interpreted as an element of $HS(K^*, H^*)$ when it acts on K^* on the right and as an element of $HS(H, K)$ when it acts on H on the left; and finally a sum of terms $\langle k|\,\langle h|$ is interpreted as an element of $HS(H, K^*)$ acting on H on the left.

We have therefore the following scheme:

$\|k\rangle\,\|h\rangle$	$HS(K^*, H)$	action on the right
$\|k\rangle\,\langle h\|$	$HS(K^*, H^*)$	action on the right
$\|k\rangle\,\langle h\|$	$HS(H, K)$	action on the left
$\langle k\|\,\langle h\|$	$HS(H, K^*)$	action on the left

The rules of calculation of adjoints can be summarized by saying that when a term consists of two factors of different types the adjoint is obtained by taking the conjugates in reverse order, while when it consists of two factors of the same type the conjugates in the same order must be considered. The transpose is obtained by exchanging left and right way of operating when a term consists of two factors of different types and by exchanging the order of factors when they are of the same type.

We now formulate the composition rules of morphisms in this notation. We consider only monomials, the general case follows by distributivity. If $L, M, and N$ are Hilbert spaces and $|m\rangle_M \;_L\langle l| \in HS(L, M)$, $|n\rangle_N \;_M\langle m'| \in HS(M, N)$, the composition of the two morphisms is

$$|n\rangle_N \;_M\langle m'| \circ |m\rangle_M \;_L\langle l| = |n\rangle_N \;_M\langle m'|m\rangle_M \;_L\langle l| .$$

If we replace M with M^* we get

$$|m'\rangle\, |n\rangle_N \circ \langle m| \;_L\langle l| = |n\rangle_N \langle m|m'\rangle \;_L\langle l| .$$

We have therefore the eight possibilities:

$$|n\rangle \langle m'| \circ |m\rangle \langle l| = |n\rangle \langle m'|m\rangle \langle l| ,$$

$$\langle n| \langle m'| \circ |m\rangle \langle l| = \langle n| \langle m'|m\rangle \langle l| ,$$

$$|n\rangle \langle m'| \circ |l\rangle\, |m\rangle = |l\rangle \langle m'|m\rangle\, |n\rangle ,$$

$$\langle n| \langle m'| \circ |l\rangle\, |m\rangle = |l\rangle \langle m'|m\rangle \langle n| ,$$

$$|m'\rangle\, |n\rangle \circ \langle m| \langle l| = |n\rangle \langle m|m'\rangle \langle l| ,$$

$$|m'\rangle \langle n| \circ \langle m| \langle l| = \langle n| \langle m|m'\rangle \langle l| ,$$

$$|m'\rangle\, |n\rangle \circ |l\rangle \langle m| = |l\rangle \langle m|m'\rangle\, |n\rangle ,$$

$$|m'\rangle \langle n| \circ |l\rangle \langle m| = |l\rangle \langle m|m'\rangle \langle n| .$$

We observe that in this notation the canonical isomorphism between $H \otimes K$ and $HS(H^*, K)$ is expressed in a very simple way for indecomposable elements:

$$|h\rangle \otimes |k\rangle \mapsto |h\rangle\, |k\rangle ,$$

and it is extended by linearity to the whole space.

In what follows we will need the following.

Lemma D.4 *If $\varphi \in HS(H^*, K)$, then it is indecomposable if and only if the image of $\varphi^* \circ \varphi$ is one-dimensional; this image is the orthogonal complement of the null space of φ.*

Proof If $\varphi = |h\rangle\, |k\rangle$, then $\varphi^* \circ \varphi = |h\rangle \langle h| \langle k|k\rangle$ and the image of $\varphi^* \circ \varphi$ is generated by $\langle h|$ so that it is one-dimensional and it is the orthogonal complement of the null space of φ. Conversely, suppose that the image of $\varphi^* \circ \varphi$ is one-dimensional. Then $\varphi^* \circ \varphi = |l\rangle \langle h|$ where $|l\rangle$ is some ket of H and $\langle h|$ some bra of H. As $\varphi^* \circ \varphi$ is self-adjoint, $\langle l|$ is collinear with $\langle h|$ and the null space of φ coincides with the null space of $\varphi^* \circ \varphi$, that is, with the orthogonal complement of $\operatorname{Im} \varphi^* \circ \varphi$. Therefore, the null space of φ is one-codimensional so that φ is indecomposable. □

The unitary group associated with a Hilbert space H is the group of linear isometries of H and will be denoted $U(H)$. It is the group of all the transformations preserving the linear structure and the scalar product of H. If X is a closed subspace of H, we can choose in it an orthonormal basis and extend it to the whole H. If Y is another closed subspace of H and there is an isometry between X and Y, by means of it we can transfer to Y the orthonormal basis in X and extend it to the whole H. As $U(H)$ is transitive on orthonormal bases, there is $T \in U(H)$ that brings from X to Y. We have therefore:

Lemma D.5 *If X and Y are closed subspaces of a Hilbert space H, every isometry between X and Y can be extended to the whole H. In particular, there is an isometry of H that brings from X to Y.*

We now prove the following theorem:

Theorem D.7 *If φ and φ' belong to $HS(H, K)$, there is $T \in U(K)$ such that $\varphi' = T\varphi$ if and only if $\varphi^*\varphi = \varphi'^*\varphi'$.*

Proof If $\varphi' = T\varphi$, we have $\varphi'^*\varphi' = \varphi^* T^* T \varphi = \varphi^*\varphi$. Vice versa, suppose that $\varphi^*\varphi = \varphi'^*\varphi'$. We first show that $\ker\varphi = \ker\varphi'$. Indeed,

$$|x\rangle \in \ker\varphi \Leftrightarrow \varphi\,|x\rangle = 0 \Leftrightarrow \langle x|\,\varphi^*\varphi\,|x\rangle = 0 \Leftrightarrow \langle x|\,\varphi'^*\varphi'\,|x\rangle = 0 \Leftrightarrow |x\rangle \in \ker\varphi'.$$

Now, using canonical factorizations and taking into account that $\pi = \pi'$, we can write

$$\pi^*\vartheta'^*\iota'^*\iota'\vartheta'\pi = \pi^*\vartheta'^*\vartheta'\pi = \pi^*\vartheta^*\vartheta\pi$$

Observing that π^* is injective, it can be canceled on the left, while π is surjective and can be canceled on the right; we conclude that $\vartheta'^*\vartheta' = \vartheta^*\vartheta$. As ϑ and ϑ' are isomorphisms, there is a unique κ such that $\vartheta' = \kappa\vartheta$ and it is expressed as $\vartheta'\vartheta^{-1}$. The isomorphism κ brings from $\operatorname{Im}\vartheta = \operatorname{Im}\varphi$ to $\operatorname{Im}\vartheta' = \operatorname{Im}\varphi'$. Furthermore, we have

$$\kappa^*\kappa = \left(\vartheta^{-1}\right)^*\vartheta'^*\vartheta'\vartheta^{-1} = \left(\vartheta^{-1}\right)^*\vartheta^*\vartheta\vartheta^{-1} = Id_{\operatorname{Im}\vartheta}.$$

Therefore, κ is an isometry between $\operatorname{Im}\varphi$ and $\operatorname{Im}\varphi'$, which can be extended to the whole K. If T is such an extension, starting from an arbitrary $|x\rangle \in \operatorname{Im}\varphi$, we apply to it the insertion ι then T. The result is the same as that obtained by applying to $|x\rangle$ first κ and then inserting the outcome in K through ι'. We conclude that $T\iota = \iota'\kappa$. Finally, we get $\varphi' = \iota'\vartheta'\pi = \iota'\kappa\vartheta\pi = T\iota\vartheta\pi = T\varphi, T \in U(K)$. □

For the sake of simplicity, we suppose henceforth that H is finite-dimensional.

If L is a Hilbert space, the group which preserves its structure is $U(L)$. But if some further structure is introduced in it, its structure group is reduced to some subgroup. Let us consider this situation when L has the structure of a tensor product. If L is (or is regarded as) a tensor product $H \otimes K$ of two Hilbert spaces H and K, we can exploit the canonical isomorphism between $H \otimes K$ and $Hom(H^*, K)$.

If we put $L = Hom(M, K)$, the groups involved are $U(M)$, $U(K)$, and $U(L)$. A triplet $(g, g', g'') \in U(M) \times U(K) \times U(L)$ preserves the structure of L as $Hom(M, K)$ if $\forall\, |x\rangle \in M, |y\rangle \in K, \varphi \in L \;\; \| y\rangle = \varphi\, |x\rangle$, then $g'\, |y\rangle = (g''\varphi)\, g\, |x\rangle$; in particular, if $\varphi = |k\rangle \langle m|$, we have $g'\, |k\rangle \langle m|x\rangle = \left[g''\, (|k\rangle \langle m|)\right] g\, |x\rangle$ whence $g'\, |k\rangle \langle m| = \left[g''\, (|k\rangle \langle m|)\right] g$, that is, $g''\, (|k\rangle \langle m|) = g'\, |k\rangle \langle m|\, g^*$.

Through the latter equation a transformation g'' arises which is well defined on the indecomposable elements of L. Indeed, for a given element, its first factor is defined up to a coefficient by whose reciprocal we must multiply the second factor, so that the right side of the equation is unaffected. But there is a canonical isomorphism between L and $M^* \otimes K$. In this isomorphism, the element $|m\rangle_{M^*} \otimes |k\rangle_K$ is sent in $|m\rangle_{M^*} |k\rangle_K$ of $Hom\left(M^{**}, K\right) = Hom(M, K)$. But $|m\rangle_{M^*} |k\rangle_K = |k\rangle \langle m|$. Then we can transfer the action of g'' to the set of indecomposable elements of $M^* \otimes K$, and uniquely extend it by linearity to the whole $M^* \otimes K$ and transfer it back to $Hom(M, K)$. Therefore, a homomorphism $U(M) \times U(K) \xrightarrow{\vartheta} U(L)$ is obtained. Its kernel is the subgroup Δ of $U(M) \times U(K) : \Delta = \left\{\left(e^{i\varphi}, e^{i\varphi}\right), \varphi \in [0, 2\pi)\right\}$. We then conclude with the following.

Theorem D.8 *If $L = Hom(M, K)$, its structure group is uniquely defined by its action on the indecomposable elements which can be expressed as $|k\rangle \langle m| \mapsto g'\, |k\rangle \langle m|\, g^*$ for some $g \in U(M)$ and some $g' \in U(K)$; this group is a subgroup of $U(L)$ and is isomorphic to $U(M) \times U(K)/\Delta$.*

It is easy to find an explicit expression for the transformations of the above group. For $g'' = (g, g') \in U(M) \times U(K)$, we put $g''\varphi = g'\varphi g^*$. Indeed, the two actions are the same on indecomposable elements, so they coincide on the whole space.

Let H be a Hilbert space. If $|x\rangle_{H^*} \in H^*$ and g is an element of $U(H)$, the position $|x\rangle_{H^*} \mapsto {}_{H^*}\langle x| = |x\rangle_H \mapsto g|x\rangle_H \mapsto {}_H\langle x|\, g^* = \widetilde{g}|x\rangle_{H^*}$ defines a canonical isomorphism between $U(H)$ and $U(H^*)$. Therefore, if $M = H^*$ and $\varphi \in Hom(H^*, K)$, we have $g''\varphi = g'\varphi\widetilde{g}^*$. But $|x\rangle_{H^*} \xrightarrow{\widetilde{g}} \widetilde{g}|x\rangle_{H^*}$ is the same as ${}_H\langle x| \mapsto {}_H\langle x|\, g^*$. Consequently, the action $|x\rangle_{H^*} \xrightarrow{\widetilde{g}^*} \widetilde{g}^*|x\rangle_{H^*}$ is the same as ${}_H\langle x| \mapsto {}_H\langle x|\, g$. If $\varphi = |h\rangle\, |k\rangle$, we have $\langle x| \mapsto \langle x|\, g\, |h\rangle\, |k\rangle \mapsto \langle x|\, g\, |h\rangle\, g'\, |k\rangle$, so that $g''\, (|h\rangle\, |k\rangle) = g\, |h\rangle\, g'\, |k\rangle$.

We have thus proved the theorem.

Theorem D.9 *If $L = Hom(H^*, K)$, its structure group is uniquely defined by its action on the indecomposable elements which can be expressed as $|h\rangle\, |k\rangle \mapsto g\, |h\rangle\, g'\, |k\rangle$ for some $g \in U(H)$ and some $g' \in U(K)$; this group is a subgroup of $U(L)$ and is isomorphic to $U(H) \times U(K)/\Delta$.*

Observing that the canonical isomorphism between $H \otimes K$ and $Hom\,(H^*, K)$ sends from indecomposable elements to indecomposable elements, we conclude with the following.

Theorem D.10 *The structure group of $H \otimes K$ is uniquely defined by its action on the indecomposable elements which can be expressed as $|h\rangle \otimes |k\rangle \mapsto g\,|h\rangle \otimes g'\,|k\rangle$ for some $g \in U\,(H)$ and some $g' \in U\,(K)$; this group is a subgroup of $U\,(H \otimes K)$ and is isomorphic to $U\,(H) \times U\,(K)/\Delta$.*

Let L denote the Hilbert space $Hom\,(H^*, K)$. In what follows it will be useful to find the subgroup $\mathscr{T}$ of $U\,(L)$ which preserves $\varphi^*\varphi$, $\varphi \in L$. We already know that $\varphi^*\varphi = \varphi^{*\prime}\varphi'$ is the necessary and sufficient condition for the existence of a transformation $T \in U\,(K)$ such that $\varphi' = T\varphi$. The scalar product in L is defined by $\langle\psi|\varphi\rangle = tr\,(\psi^*\varphi)$, so that the above position defines an element of $U\,(L)$ and consequently a natural homomorphism from $U\,(K)$ to $U\,(L)$. But other transformations with this property may exist in $U\,(L)$ not lying in $I \times U\,(K)$. Therefore a detailed investigation is required. We first observe that the scalar product in L is $tr\,(\psi^*\varphi)$. If φ is considered as an element of a Hilbert space, it must be regarded as a ket $|\varphi\rangle$, and the same thing for ψ. Hence we must put $\langle\psi| = tr\,(\psi^*\circ)$. In particular, if $\psi = |h\rangle\,|k\rangle$ and $\varphi = |x\rangle\,|y\rangle$, $\langle\psi|\varphi\rangle = tr\,(\langle h|\,\langle k| \circ |x\rangle\,|y\rangle) = \langle h|x\rangle\,\langle k|y\rangle$. Therefore, if $|\psi\rangle = ||h\rangle\,|k\rangle\rangle$, then $\langle\psi| = \langle\langle h|\,\langle k||$.

Let $g \in U\,(L)$, $\varphi = |h\rangle\,|k\rangle$ and $|\varphi'\rangle = g\,|\varphi\rangle$. We have $\varphi^*\varphi = \langle h|\,\langle k| \circ |h\rangle\,|k\rangle = |h\rangle\,\langle h|\,\langle k|k\rangle = \varphi'^*\varphi'$. Hence, by Lemma D.4, $\varphi' = |h\rangle\,|k'\rangle$, $|k'\rangle$ depends linearly on $|k\rangle$ and $\langle k|k\rangle = \langle k'|k'\rangle$. We conclude that g is in $I \times U\,(K)$ and that the latter is the group of all the unitary transformations preserving $\varphi^*\varphi$. Therefore we have proved the following.

Theorem D.11 *If $L = Hom\,(H^*, K)$, the subgroup of $U\,(L)$ which preserves $\varphi^*\varphi$ for every $\varphi \in L$ is uniquely defined by its action on the indecomposable $I \times U\,(K)$.*

D.3.1 Orbits. Spectra

Let us now consider the orbits of vectors of L under the action of its structure group (as $Hom\,(H^*, L)$). Starting from a vector φ we can act on it first with $U\,(K)$ generating the orbit under this action and then with $U\,(H)$ in order to obtain the full orbit. We observe that, as the actions of $U\,(H)$ and $U\,(K)$ are permutable, the full orbit is the union of the orbits relative to $U\,(K)$. Each orbit relative to the latter action is labeled with the operator $\rho = \varphi^*\varphi$; the action of $U\,(H)$ on it is effected by orthogonal similitude transformations. We conclude that each full orbit can be characterized by the spectrum of the similitude class.

Remembering that ρ is a positive operator, we can describe the spectrum as a mapping $\mathfrak{S}$ from $\mathbb{R}_+$ to $\mathbb{N}_0$ almost everywhere zero. The operator ρ itself can be represented by its spectral decomposition, which in turn can be described as a mapping $\mathfrak{D}$ from $\mathbb{R}_+$ to the set of subspaces of H^* almost everywhere zero and such that H^* is their orthogonal direct sum. If $\mathfrak{N}$ is the mapping which associates to each subspace of H its dimension, we have $\mathfrak{S} = \mathfrak{N} \circ \mathfrak{D}$. The connection between ρ and $\mathfrak{D}$ is the following:

$$\rho = \sum_{\lambda \in \mathbb{R}_+} P(\mathfrak{D}(\lambda))\lambda,$$

where $P(X)$ denotes the projector on X.

We can restrict each spectrum $\mathfrak{S}$ to its support and to its image, and the same thing we will do with $\mathfrak{D}$. With an abuse of notation, we will use the same symbols for these restrictions and, if not differently specified, we will refer implicitly to the restrictions. Furthermore, we will call spectrum any surjective mapping from a subset of $\mathbb{R}_+$ to a subset of $\mathbb{N}_0$.

The mapping Π which associates each spectral decomposition $\mathfrak{D}$ with the corresponding spectrum $\mathfrak{S}$ is clearly surjective, so that we can regard the set $\mathfrak{D}$ of all spectral decompositions as a fibered space. Each fiber corresponds bijectively to an orbit of the action of the structure group $U(H^*) \times U(K)/\Delta$ of the tensor product on L. Furthermore, we can define a mapping Θ which associates to each spectral decomposition $\mathfrak{D}$ its image $\operatorname{Im} \mathfrak{D}$ which is nothing but a decomposition of H^* in an orthogonal direct sum. Henceforth, if not differently specified, by decomposition of a Hilbert space we mean a decomposition of it in an orthogonal direct sum.

With this terminology, the image of Θ is the set $\mathscr{C}$ of all possible decompositions of H^*. Denoting with $\mathscr{S}$ the set of all possible spectra, we have the diagram of mappings $\mathscr{C} \overset{\Theta}{\leftarrow} \mathscr{D} \overset{\Pi}{\rightarrow} \mathscr{S}$. We emphasize that, although the mappings are surjective, they do not define a direct product because the elements of $\mathscr{C}$ and of $\mathscr{S}$ cannot be chosen independently and, even when they are compatible, they do not define a unique element of $\mathscr{D}$. In order to clarify this point, it is convenient to introduce the following definition:

Definition D.1 We call type of a spectrum $\mathfrak{S}$ the mapping which associates to each multiplicity the number of eigenvalues having this multiplicity; we call type of a decomposition $\mathfrak{C}$ the mapping which associates to each multiplicity the number of eigenspaces having this multiplicity.

We will introduce a specific notation to indicate a type $\mathscr{T}$. The set of multiplicities appearing in $\mathscr{T}$ will be put in increasing order in a list; the number of eigenvalues (eigenspaces) associated with each multiplicity will appear as a superscript of the multiplicity. We will therefore write: $\mathscr{T} = \left\{\mu_1^{n_1} \mu_2^{n_2} ... \mu_r^{n_r}\right\}$ with $\mu_1 < \mu_2 < ... < \mu_r$. If the type refers to a spectrum, the number of different eigenvalues is $n_1 + n_2 + ... + n_r$, while, when counted with their multiplicity, their number is $n_1\mu_1 + n_2\mu_2 + ... + n_r\mu_r$. There is a constraint for the type because the latter quantity must be equal to the dimension of the space. If the type refers to a decomposition, $\mu_1, \mu_2 ..., \mu_r$ are the dimensions of the eigenspaces involved in the decomposition, $n_1, n_2, ..., n_r$ are their number of occurrences in the decomposition and $n_1\mu_1 + n_2\mu_2 + ... + n_r\mu_r$ is the dimension of the space.

Let $\mathfrak{C}$ be a decomposition of H^*, $\mathscr{H}_\mu$ the set of spaces of $\mathfrak{C}$ of dimension μ and $\mathscr{P}(\mathscr{H}_\mu)$ the group of permutations of $\mathscr{H}_\mu$. We define the group $\mathscr{G} = \prod_\mu \times \mathscr{P}(\mathscr{H}_\mu)$ where $\times$ indicates that the product must be interpreted as a group direct product.

Given $\mathfrak{C} \in \mathscr{C}$ and $\mathfrak{S} \in \mathscr{S}$, there is $\mathfrak{D} \in \mathscr{D}$ such that $\mathfrak{C} = \Theta(\mathfrak{D})$ and $\mathfrak{S} = \Pi(\mathfrak{D})$ if and only if $\mathscr{T}(\mathfrak{C}) = \mathscr{T}(\mathfrak{S})$. If the latter condition is satisfied, $\mathfrak{D}$ is defined up to the composition on the left with an arbitrary element of $\mathscr{G}$.

Indeed, for each μ the spectral decomposition $\mathfrak{D}$ establishes a bijection from the set of eigenvalues possessing this multiplicity and the set of the corresponding eigenspaces whose dimension must be equal to μ. Therefore the cardinality of these two sets must be the same, and then $\mathscr{T}(\mathfrak{C}) = \mathscr{T}(\mathfrak{S})$. Vice versa, suppose that the latter condition is satisfied. Then for each μ the inverse image of μ for the mapping $\mathfrak{S}$ and the set of spaces of dimension μ of the decomposition $\mathfrak{C}$ have the same cardinality, so that there is a bijection β from the first to the second set. We therefore obtain a bijection $\mathfrak{D}$ from the domain of $\mathfrak{S}$ to the decomposition $\mathfrak{C}$. Such a bijection is a spectral decomposition with the required property. Each bijection β is defined up to the composition on the left with an arbitrary element of $\mathscr{P}(\mathscr{H}_\mu)$. Therefore $\mathfrak{D}$ is defined up to the composition on the left with an arbitrary element of $\mathscr{G}$.

In what follows the structure group of $Hom\,(H^*, K)$ will be denoted $\mathfrak{H}$.

The orbits of $U\,(L)$ in the Hilbert space L are the loci of constant norm. The effect of the group reduction from $U\,(L)$ to $\mathfrak{H}$ is to split each orbit in a set of orbits under the action of $\mathfrak{H}$. Each of the latter orbits is labeled by a spectrum $\mathfrak{S} \in \mathscr{S}$. So, while the orbits under $U\,(L)$ are labeled by squared norms, the orbits under $\mathfrak{H}$ are labeled by spectra. A squared norm is $\langle\varphi|\varphi\rangle = tr\,(\varphi^*\varphi) = \sum_p p\,\mathfrak{S}\,(p)$. The space L as a tensor product can be regarded as a fibered space with base $\mathscr{S}$ and whose projection Ψ is the composition of the mapping Ξ which associates to each φ the spectral decomposition of $\varphi^*\varphi$ with the mapping Π previously introduced, i.e., $\Psi = \Pi \circ \Xi$. A fiber of this space is a $\mathfrak{H}$-orbit.

The spectra can be classified according to their type, so that also the fibers are classified accordingly. Each fiber is a Homogeneous $\mathfrak{H}$-space. Hence it can be characterized by means of its conjugation class of stabilizers. In order to investigate this question it is useful to introduce the Schmidt representation of φ (which is nothing but the abstract version of the singular value decomposition of a matrix).

Let $\langle x|$ be an eigenbra of $\varphi^*\varphi$. Then $\varphi^*\,(\langle x|\,\varphi) = \langle x|\,p$ for some $p \in Dom(\mathfrak{S})$. Then $(\varphi\varphi^*)\,(\langle x|\,\varphi) = \varphi\,(\langle x|\,p) = \langle x|\,\varphi p$. Consequently, if $\langle x|$ does not belong to the null space of φ, $\langle x|\,\varphi$ is an eigenket of $\varphi\varphi^*$ with eigenvalue p. But $\langle x|$ belongs to the null space of φ if and only if $\langle x|\,\varphi\varphi^* = 0$, that is $\varphi^*\,(\langle x|\,\varphi) = 0$, i.e., if and only if $p = 0$. We conclude that if $\langle x|$ is an eigenbra of $\varphi^*\varphi$ with nonzero eigenvalue, $\langle x|\,\varphi$ is an eigenket of $\varphi\varphi^*$ with the same eigenvalue. As already observed, the mapping $\mathfrak{R} \xrightarrow{\Pi} \mathscr{S}$ defines a structure of fibered space. The classification of the spectra induces a classification of the fibers, so that we can speak of type of a fiber as the type of its projection on the base. The fibered space can therefore be decomposed into the disjoint union of fibered subspaces; in each of them all the fibers are of the same type and we can introduce the notion of type for such subspaces.

Let us concentrate our attention to a subspace of a given type $\mathscr{P}$. An element $\mathfrak{S}$ of its base can be identified by a set of $m = n_1 + n_2 + ... + n_r$ real nonnegative numbers which will be represented in increasing order in a list $\mathfrak{P} = \{p_1, p_2, ..., p_m\}$.

With the terminology introduced above, the compatibility condition can be expressed saying that the necessary and sufficient condition for the compatibility of $(\mathfrak{C}, \mathfrak{S})$ is that $\mathscr{T}(\mathfrak{C}) = \mathscr{T}(\mathfrak{S})$. For a given $\mathscr{T}$, $\mathfrak{C}$ is specified by giving n_1 spaces of dimension μ_1, n_2 of dimension μ_2 and so on; hence we can indicate it with the symbol $\mathfrak{C} = \left\{H^{*\ n_1}_{\mu_1} H^{*\ n_2}_{\mu_2} ... H^{*\ n_r}_{\mu_r}\right\}$ where it is understood that all products are orthogonal direct products. As the compatibility condition consists simply in the identification of types, for a given type $\mathfrak{C}$ and $\mathfrak{S}$ can be assigned independently. As we are considering all the objects relative to a fixed type $\mathscr{T}$, with an abuse of notation we continue to denote with $\mathfrak{C}$ and $\mathfrak{S}$ the sets of decompositions and of spectra belonging to that type. On the other hand, the specification of $\mathfrak{C}$ and $\mathfrak{S}$ defines ρ up to a transformation of $\mathscr{G}$. We will say that ρ is of type $\mathscr{T}$ if its spectrum is of type $\mathscr{T}$. With the same abuse of notation, we continue to denote with $\mathfrak{R}$ the set of density operators of type $\mathscr{T}$. Restricting ourselves to this type, we can consider $\mathfrak{R}$ as a $\mathscr{G}$-space whose orbits are specified by elements of $\mathscr{C} \times \mathscr{S}$. We can classify the elements of L according to the type of the corresponding $\mathfrak{S}$. If we limit ourselves to the elements of L of type $\mathscr{T}$ and continue to call L the set of such elements (which is not a linear space), we can specify the orbits of the action on L of the structure group of the tensor product by the assignment of $\mathfrak{R}$. If we restrict the structure group to $U(K)$, the orbits of this action are specified by the assignment of an element of $\mathfrak{R}$. The latter amounts to specify, besides $\mathfrak{P}$, an element $\mathfrak{C} \in \mathscr{C}$.

D.4 Class of Hilbert Spaces as a Category

In what follows it will be useful to regard the class of Hilbert spaces as a category. The objects of this category are the Hilbert spaces themselves, and the morphisms are the bounded linear maps between them. In this category, there is an involution which associates an object with its adjoint and a morphism with its adjoint. Owing to the rule $(\varphi \circ \psi)^* = \psi^* \circ \varphi^*$, the involution is a contravariant functor between $\mathscr{H}$ and itself.

Definition D.2 A G-space is a doublet (G, X) where X is a set and G a group of its transformations. Note that in our definition G is not an abstract group, but a group of transformations of X.

We want to introduce the concept of a category of G-spaces. To this purpose, we have to introduce the definition of morphism of G-spaces.

Definition D.3 A morphism $(G, X) \xrightarrow{\gamma} (K, Y)$ is a doublet (α, φ) where $X \xrightarrow{\varphi} Y$ is a mapping and $G \xrightarrow{\alpha} K$ is a group morphism such that $\varphi(gx) = \alpha(g)\varphi(x)\, \forall x \in X, g \in G$.

It can be easily shown that, defining $\left(\alpha', \varphi'\right)(\alpha, \varphi) = \left(\alpha' \circ \alpha, \varphi' \circ \varphi\right)$, we get a category.

We define in X the equivalence relation: $x \sim x' \Leftrightarrow \varphi(x) = \varphi\left(x'\right)$. We can show that: $x \sim x' \Rightarrow gx \sim gx'$. Indeed $\varphi(gx) = \alpha(g)\varphi(x) = \alpha(g)\varphi\left(x'\right) = \varphi\left(gx'\right)$, so that $gx \sim gx'$.

If $Z = X/\sim$, and π is the canonical projection from X to Z, we can associate with π a unique group morphism β such that (β, π) is a G-space morphism according to the following.

Theorem D.12 *If (G, X) is a G-space and $X \xrightarrow{\pi} W$ is a surjective mapping such that $\pi(x) = \pi\left(x'\right) \Rightarrow \pi(gx) = \pi\left(gx'\right)$, then there is a unique group morphism ν such that (ν, π) is a G-space morphism and that for every morphism of the form $(G, X) \xrightarrow{(\mu,\pi)} \left(G', W\right)$ μ uniquely factorizes through ν on the right.*

Proof We show first that ν exists. If $w \in W$, there is $x \in X$ such that $w = \pi(x)$. We put $\nu(g)(w) = \pi(gx)$. The definition is well posed. Indeed, if $w = \pi\left(x'\right)$, $\pi(x) = \pi\left(x'\right)$ and then $\pi(gx) = \pi\left(gx'\right)$. We have

$$\nu\left(gg'\right)(w) = \pi\left(gg'x\right) = \pi\left(g\left(g'x\right)\right) = \nu(g)\pi\left(g'x\right) = \nu(g)\nu\left(g'\right)\pi(x) = \nu(g)\nu\left(g'\right)(w)$$

so that $\nu\left(gg'\right) = \nu(g)\nu\left(g'\right)$. This equation shows that each $\nu(g)$ is a transformation of W and that ν is a group morphism. If $\operatorname{Im}\nu = K$ and, with an abuse of notation, we continue to denote ν its restriction to K, (ν, π) is a morphism $(G, X) \xrightarrow{(\nu,\pi)} (K, Y)$. If $(G, X) \xrightarrow{(\mu,\pi)} \left(G', W\right)$, we must have $\mu(g)(w) = \mu(g)\pi(x) = \pi(gx) = \nu(g)(w)$, so that $G' \supseteq \operatorname{Im}\mu = \operatorname{Im}\nu = K$, and $\mu = \iota \circ \nu$, where ι is the insertion of K in G'. As ν is surjective, it is right-cancellable, so that the factorization through ν is unique. If ν' also satisfies all the conditions of the theorem, $\operatorname{Im}\nu' = \operatorname{Im}\nu$ and $\nu' = \iota \circ \nu$, so that $\iota = id_K$ and $\nu' = \nu$. □

D.5 System and Environment

Let us consider a system described by a Hilbert space H interacting with "the rest of Universe," represented by the Hilbert space K. In what follows the first system will be called simply the system, while the second system will be referred as the environment.

According to standard Quantum Mechanics, the states of the composite system are the rays of the tensor product $H \otimes K$, while the states of the system isolated from the environment are the rays of H. Our next task is to give a natural characterization of the states of the system considered as open, that is, taking into account the presence of the environment through a minimal set of data. It's exactly what we do for a classical open system, where the effect of the environment is represented by a minimal set of inputs. In order to accomplish this task, it is useful to see Hilbert spaces as G-spaces, i.e., as sets equipped with a selected group of allowed transformations.

The Hilbert space of the environment is K over which an action of the unitary group $U(K)$ is defined. The group $I \times U(K)$ acts on $H \otimes K$ in such a way so as to leave unchanged the vectors of H. This group induces an action on the set of rays of $H \otimes K$, that is, on the set of states of the composite system, leaving unchanged the states of the system (as a closed system). Furthermore, the group acts transitively on the states of the environment (as a closed system), so that every state of the environment can be reached from any other by means of transformations of the group.

It is reasonable to say that a property of a state of the composite system is a property of the open system alone if and only if it remains unchanged under an arbitrary change of the state of the environment because such a property is independent on the specific state of the environment, but at the same time the existence of it is not ignored. On the basis of the above discussion, we deduce that such a property is a function uniquely of the orbit described by a state of the composite system under the action of the group $I \times \mathfrak{A}(K)$. We finally arrive in a natural way to the following.

Definition D.4 A state of the open system is an orbit of a state of the composite system under the action of the group $I \times \mathfrak{A}(K)$.

Although the above definition captures the essential features of the state of the open system (i.e., that the state of the open system is something independent of the state of the environment but at the same time it takes into account its presence), mention of the states of the composite system is involved in it.

In order to eliminate from the definition any reference to the states of the composite system, we proceed as follows. We will find a map from the set of states of the composite system onto a set of mathematical objects built up uniquely in terms of H such that two states of the composite system belong to the same orbit if and only if their images through the map are equal. In this way, we obtain a bijection between the orbits and such objects. Therefore, we can use these objects to label the states of the open system and any reference to the states of the composite system is eliminated.

Owing to the canonical isomorphisms previously introduced, we can regard the states of the composite system as rays of $Hom(H^*, K)$. The action of an element $(I, T) \in I \times U(K)$ on $H \otimes K$ corresponds to the action $\varphi \mapsto T\varphi(\varphi \in Hom(H^*, K))$. On the basis of Theorem D.7 we easily conclude the following.

If two states of the composite system are represented by φ and φ', the necessary and sufficient condition for them to belong to the same orbit is that $\varphi'^*\varphi'$ and $\varphi^*\varphi$ are different only by a numerical factor.

Choosing orthonormal bases in H and K, it is easily shown that $\|\varphi\|^2 = tr\varphi^*\varphi$. If we normalize to 1 the state vectors of the composite system, we conclude that each orbit (i.e., each state of the open system) is characterized by a positive trace class operator, normalized to 1.

In the conventional approach, where the probabilistic interpretation and the associated Born rule are postulated a priori, such an operator is a density operator and it describes a mixture of pure states. We emphasize that instead here no probability interpretation has as yet been introduced.

Here on we will suppose, for the sake of simplicity, that the system is described by a finite-dimensional Hilbert space, while the space of the environment will be supposed infinite-dimensional.

Conversely, suppose that a density operator ρ for the system is prescribed. Its spectral decomposition yields a finite set of eigenspaces to each of which a nonnegative eigenvalue is associated. These eigenvalues sum to 1. We choose an orthonormal basis in each eigenspace, so that we obtain a finite sequence $|u_1\rangle, |u_2\rangle, \ldots |u_n\rangle$ with their corresponding eigenvalues $p_1, p_2, \ldots p_n$ and ρ is expressed as $\rho = \sum_{i=1}^{n} p_i |u_i\rangle \langle u_i|$. Correspondingly, we choose in K a set of orthonormal vectors $|v_1\rangle, |v_2\rangle, \ldots |v_n\rangle$, and define $\varphi \in Hom(H^*, K)$ as the unique morphism in which $\langle u_1| \mapsto \sqrt{p_1}|v_1\rangle, \langle u_2| \mapsto \sqrt{p_2}|v_2\rangle, \ldots \langle u_n| \mapsto \sqrt{p_n}|v_n\rangle$. This morphism is expressed by $\varphi = \sum_{i=1}^{n} \sqrt{p_i}|u_i\rangle |v_i\rangle$. The adjoint morphism is $\varphi^* = \sum_{i=1}^{n} \sqrt{p_i}\langle u_i| \langle v_i|$. But $(\langle u_i| \langle v_i|) \circ (|u_j\rangle |v_j\rangle) = |u_j\rangle \delta_{ji} \langle u_i|$, so that $\varphi^*\varphi = \sum_{i=1}^{n} p_i |u_i\rangle \langle u_i| = \rho$.

We conclude that *for each density operator ρ there is a state of the composite system φ such that $\rho = \varphi^*\varphi$.*

On the basis of the spectral decomposition of ρ, the state of the open system is completely specified by the data of its spectral decomposition, i.e., by the list of the eigenspaces, each associated with the corresponding eigenvalue. For the moment, we suppose that the Hilbert space of the system is two-dimensional, so that the system is, for example, a spin. We further suppose to consider a state whose density operator is nondegenerate, so that each eigenvalue is different from $1/2$. Therefore, a state will be described by the set $\{(s_1, p_1), (s_2, p_2)\}$ where s_1, s_2 are rays of H^* and p_1, p_2 their corresponding eigenvalues (such that $p_1 + p_2 = 1$). So, a state of the open system is different from any state of the closed one: it is a weighted set of states, in the sense that to each member of the set a weight is attributed. Our next goal is to investigate the nature of these weights.

Suppose that ρ derives from a state of the composite system represented by a normalized $\varphi \in Hom(H^*, K)$. If $\langle u_1|, \langle u_2|$ represent the rays of s_1, s_2, respectively, their images through φ will be, say, $|v_1\rangle, |v_2\rangle$. If $\langle u_1|, \langle u_2|$ are normalized, the element φ is expressed as $\varphi = |u_1\rangle |v_1\rangle + |u_2\rangle |v_2\rangle$. Its adjoint is $\varphi^* = \langle u_1| \langle v_1| + \langle u_2| \langle v_2|$. The linear operator ρ acts on the rays of H. It generates a semigroup, to which a discrete-time dynamics is associated. The equilibrium points of this dynamics are the eigenstates of ρ, and they are exactly two as long as ρ is nondegenerate. The eigenstate associated with the eigenvalue $> 1/2$ is stable, while the other one is unstable. If we consider the set of all dynamics with the same equilibrium points but with different nondegenerate spectra, this set can be partitioned into two classes, such that two elements belong to the same class if their stable point is the same. Correspondingly, we have a set of orbits partitioned into two classes.

We will now show that a *measure* can be introduced in a natural way in the union set of this set.

Let us first introduce some further notation. The set of rays of a complex linear space is called a complex projective space, and the rays are called points. If a closed quantum system is represented through a Hilbert space H, its states are rays of H as a linear space, that is, points of the associate projective space, which will be denoted $\mathscr{P}(H)$. For sake of brevity, we put $L = Hom\,(H^*, K)$. The states of the composite system are therefore the points of $\mathscr{P}(L)$.

If is a state of the (open) system, $|u_1\rangle, |u_2\rangle$ are normalized eigenvectors of ρ and p_1, p_2 the corresponding eigenvalues, a possible state of the composite system compatible with ρ is represented by $\varphi = |u_1\rangle|v_1\rangle + |u_2\rangle|v_2\rangle$, where $\langle v_1|v_1\rangle = p_1$, $\langle v_2|v_2\rangle = p_2$, $\langle v_1|v_2\rangle = 0$. For given φ, $|v_1\rangle$ and $|v_2\rangle$ are defined up to arbitrary phase factors. If we choose another representative of the same state of the composite system, a further phase factor common to both vector arises, but the latter can be absorbed in the former. In addition, when the whole orbit is generated operating on through $U(K)$, the phase factors can be absorbed in its transformations.

There is still an ambiguity to be dealt with. We have not yet specified which eigenvector is labeled with 1 and which with 2. As $p_1 \neq p_2$, we will label with 1 the greater eigenvalue and the corresponding eigenvector. We conclude that, given ρ, there is a doublet $(|v_1\rangle, |v_2\rangle)$ of kets of K such that $\langle v_1|v_1\rangle = p_1$, $\langle v_2|v_2\rangle = p_2$, $\langle v_1|v_2\rangle = 0$, $\varphi = |u_1\rangle|v_1\rangle + |u_2\rangle|v_2\rangle$, and $\rho = \varphi^*\varphi$.

As long as $p_1 p_2 = 0$, a two-dimensional subspace S of K is defined.

Let $\left(|v'_1\rangle, |v'_2\rangle\right)$ be another doublet which generates a subspace S'. We want to establish when it defines the same ρ as $(|v_1\rangle, |v_2\rangle)$. This happens if and only if there is $T \in U(K)$ such that $|v'_1\rangle = T|v_1\rangle$ and $|v'_2\rangle = T|v_2\rangle$. Such a transformation exists if and only if there is an isometry between S and S' which sends $|v_1\rangle$ in $|v'_1\rangle$ and $|v_2\rangle$ in $|v'_2\rangle$. We conclude that they define the same ρ if and only if they are also orthogonal and with the same norms.

The space L is a linear space which, when equipped with the scalar product $tr\psi^*\varphi$, becomes a Hilbert space. But in L we can also define $\psi^*\varphi$ with values in $Hom\,(H^*, H^*)$. The group of linear transformations preserving the scalar product is $U(L)$, while the group Γ of linear transformations preserving $\psi^*\varphi$ is obviously a subgroup of it.

We now define formally the category of Hilbert spaces. Its objects are linear spaces equipped with a scalar product, which are complete with respect to the induced norm. Its morphisms are all the linear maps which are continuous with respect to this norm. To each morphism φ, we can associate its adjoint φ^*. We have the rules $(\varphi + \psi)^* = \varphi^* + \psi^*$; $(\varphi\psi)^* = \psi^*\varphi^*$; $(\lambda\varphi)^* = \lambda^*\varphi^*$; $(\varphi^*)^* = \varphi$. The operation of passage to the adjoint is an involution. The complex field $\mathbb{C}$ has the structure of a Hilbert space in an obvious way. Each vector of a Hilbert space can be replaced uniquely by a morphism in the following way. If $x \in H$, there is a unique morphism

$\mathbb{C} \xrightarrow{\varphi} H$ of linear spaces such that $\varphi(1) = x$. Conversely, given $\mathbb{C} \xrightarrow{\varphi} H$, we define $x = \varphi(1)$. The whole H can be replaced by $Hom(\mathbb{C}, H)$. The sum of vectors can be replaced by the sum of corresponding morphisms, and so on. The scalar product $\langle x|y\rangle$ is replaced by $\varphi^* \psi$ regarded as an endomorphism of $\mathbb{C}$ (where $x = \varphi(1)$ and $y = \psi(1)$).

It is useful to introduce a notation similar to the one adopted in ordinary tensor algebra. If $x \in H$, we will write x^H. If $y \in H^*$, we will write y_H. The number $y(x)$ will be denoted $y_H \circ x^H$. The functional y_H is the adjoint of some vector y^H, so that we put $y_H = \left(y^H\right)^*$. On the other hand, y_H is a vector of H^*, and as such it will be indicated as y^{H^*}. More generally, $x^{HK...L}$ will denote an element of $H \otimes K \otimes ... \otimes L$.

We observe that a ket $|k\rangle \in K$ corresponds bijectively to a morphism $\mathbb{C} \xrightarrow{|k\rangle} K$: $c \mapsto c\,|k\rangle$. Writing the scalar on the right, we will regard this morphism as operating on the left. The same morphism can be written as acting on the right. Its transpose $\widehat{|k\rangle}$ operates according to the rule $\widehat{|k\rangle}\,\langle y| = \langle y|k\rangle$ while, if $|z\rangle = |k\rangle\, c$, $c^*|k\rangle^* = \langle z| = c^*\,\langle k|$, so that the adjoint of $|k\rangle$ as a morphism is the bra $\langle k|$. Therefore, when $|k\rangle$ operates on the left on $\mathbb{C}$ it is regarded as an element of $Hom(\mathbb{C}, K)$, while when it operates on the right on K it defines the transpose. For a bra, left and right are interchanged. Let now $|h\rangle \in H$ and $|k\rangle \in K$. If $|h\rangle$ is interpreted as an element of $Hom(\mathbb{C}, H)$, $|k\rangle$ must be interpreted as an element of $Hom(K^*, \mathbb{C})$ in order to allow a composition; if $\langle x| \in K^*$, we operate on the right on it with $|k\rangle$ and then we operate with $|h\rangle$ on the right, obtaining the composed morphism $|k\rangle\,|h\rangle \in Hom\,(K^*, H)$. Of course, the other possibility is obtained interchanging the role of $|h\rangle$ and $|k\rangle$. We have $\langle h|\,\langle k| = |h\rangle^*|k\rangle^* = (|k\rangle\,|h\rangle)^*$.

Let us write the relationship between the scalar product in H and in H^* in the bra-ket notation. We remember that $\langle x|y\rangle_{H^*} = \langle y^*|x^*\rangle_H$. But $x = |x\rangle_{H^*}$, so that $x^* = {}_{H^*}\langle x| = |x\rangle_H$. Similarly, $y^* = |y\rangle_H$. Therefore $\langle x|y\rangle_{H^*} = \langle |y\rangle_H \mid x\rangle_H\rangle = \langle y.$ Hence *the rule of calculation* $\langle x|y\rangle_{H^*} = \langle y|x\rangle_H$ holds. The equation ${}_H\langle x| = |x\rangle_{H^*}$ can be interpreted as an equation between functionals. If we apply both members to $|y\rangle_H$, we get ${}_H\langle x|y\rangle_H = |x\rangle_{H^*}|y\rangle_H = {}_{H^*}\langle y|x\rangle_{H^*} = |y\rangle_H|x\rangle_{H^*}$.

D.6 Measurement

We afford here the problem of measurement in quantum systems. A measurement is a special kind of interaction between an observed system and an apparatus. In an ideal measurement at the end of the interaction process, there is a one-to-one correspondence between a set of states $|h_i\rangle$ of the observed system and a set of states $|k_i\rangle$ of a subsystem of the apparatus which we call the pointer. Consistently, the states $|k_i\rangle$ will be referred to as the pointer states. The final state of the pointer is some density operator, which for the moment we suppose generic. The result of a specific experiment yields one among the states $|k_i\rangle$. Consequently, the latter are exactly the non-null eigenstates of the density operator representing the final state of the pointer, so that they form an orthonormal set. The correspondence is fixed as a result of the interaction, in the sense that the set of possible final states of observed system plus pointer is exactly the set $\{|h_i\rangle\, |k_i\rangle\}$. In this way, we are ensured that, if in a specific experiment the final state of the pointer is $|k_i\rangle$, the final state of the observed system is $|h_i\rangle$. An immediate consequence of the above considerations is that the final state of observed system plus pointer is a mixed state. This fact entails that the apparatus must be composed with the pointer and some other subsystem; the latter will be called the detector. With this terminology the apparatus consists of a detector and a pointer. Let $H, K, and L$ denote the Hilbert spaces of the observed system (shortly the system), the pointer, and the detector, respectively, and $W = H \otimes K \otimes L$ the Hilbert space of the whole system. We will denote with the same symbol the systems and their Hilbert spaces. Supposing that the density operator ρ_{HK} is generic, the canonical decomposition of the final state of the whole system is $|\psi\rangle = \sum d_i\, |h_i k_i\rangle\, |l_i\rangle$, where the $|l_i\rangle$ form an orthonormal set of L and the d_i are positive numbers all different from each other. The possible states of H, given that the state of the whole system W is $|\psi\rangle$ and that the state of the pointer is any of the pointer states, are the $|h_i\rangle$. On the other hand, the possible states of H, given that the state of the whole system W is $|\psi\rangle$, are the eigenstates of ρ_H. In an ideal measurement, we require that the two sets of states are the same.

D.7 Stabilizer

Let $H \otimes K$ be the tensor product of two finite-dimensional Hilbert spaces. Let $|\psi\rangle \in H \otimes K$. Let G_ψ be the stabilizer of $|\psi\rangle$ in $U(H) \otimes U(K)$.

In order to describe the structure of G_ψ we recall the canonical decomposition of $|\psi\rangle$, based on the following THEOREM (1zurekstate I), where tr_E denotes partial trace with respect to K.

Theorem *If $|\psi\rangle \in H \otimes K$, there is a unique decomposition*

$$|\psi\rangle = \sum d_\alpha\, |\chi_\alpha\rangle$$

such that $d_\alpha > 0$ with the d_α all different from each other,

$$tr_E\, |\chi_\alpha\rangle \langle \chi_\beta| = 0$$

for $\alpha \neq \beta$, and the set

$$\{P_\alpha = tr_E\, |\chi_\alpha\rangle\, \langle \chi_\alpha|\}$$

is a family of orthogonal projectors of H. Furthermore

$$tr_E\, |\psi\rangle\, \langle \psi| = \sum d_\alpha^2 P_\alpha.$$

For each $|h\rangle \in H$, the quantity

$$(\langle h| \otimes \langle x|)\, |\chi_\alpha\rangle$$

depends antilinearly on $|x\rangle$, defining a ket of K depending antilinearly on $|h\rangle$ and hence linearly on $\langle h|$. Hence $|\chi_\alpha\rangle$ uniquely defines a linear mapping from H^* to K which we call χ_α.

We introduce

$$\chi = \sum \chi_\alpha .$$

If H is the non-null eigenspace of the projector

$$P = \sum P_\alpha ,$$

the restriction of χ to H^* and to its image is an isometry η (1zurekstate I).

Let H_α be the non-null eigenspace of P_α and Υ_α the subgroup of $U(H)$ acting as identity on $H_{\alpha\perp}$.

As the groups Υ_α are mutually commutative, we can define their internal direct product

$$\Upsilon_+ = \times \Upsilon_\alpha .$$

We further define Υ_0 as the subgroup of $U(H)$ acting as the identity on H and Υ'_0 as the subgroup of $U(K)$ acting as the identity on Im χ.

There is a canonical isomorphism which associates any element of Υ_+ with its restriction to H; let $\widehat{\Upsilon}_+$ be the image of this isomorphism; defining

$$\tilde{\gamma}\, \langle h| = \langle h|\, \gamma^* ,$$

to each $\gamma \in \widehat{\Upsilon}_+$ we associate the element $\eta\tilde{\gamma}\eta^{-1}$ belonging to the unitary group of Im χ and in this way we obtain an isomorphism υ between $\widehat{\Upsilon}_+$ and a subgroup $\widehat{\Upsilon}'_+$ of the latter group.

We introduce finally the group $\widehat{\Gamma}$ as the graph of υ. The isomorphism υ induces an isomorphism $\widehat{\upsilon}$ between $\widehat{\Upsilon}_+$ and $\widehat{\Gamma}$. The group $\widehat{\Upsilon}_+$ inherits from Υ_+ the structure of a direct product, namely, if $\widehat{\Upsilon}_\alpha$ is the image of Υ_α through the canonical isomorphism, we have $\widehat{\Upsilon}_+ = \times \widehat{\Upsilon}_\alpha$. Thus $\widehat{\Gamma} = \times \widehat{\Gamma}_\alpha$ where $\widehat{\Gamma}_\alpha$ is the image of $\widehat{\Upsilon}_\alpha$ through $\widehat{\upsilon}$.

The groups $\widehat{\Gamma}_\alpha$ and $\widehat{\Gamma}$ act on $H \times$ Im χ; their action can be extended to $H \times K$ by requiring that they act as the identity on $H_\perp \times (\text{Im}\, \chi)_\perp$. We thus define the groups Γ_α and $\Gamma = \times \Gamma_\alpha$ acting on the whole $H \times K$.

If $(U, V) \in U(H) \times U(K)$, the position $(U, V) \mapsto U \otimes V$ defines an epimorphism

$$U(H) \times U(K) \xrightarrow{\theta} U(H) \otimes U(K)$$

of groups. The kernel of this morphism is the group

$$\left\{\left(cI, c^{*}I\right)|c \in \mathbb{C}_1\right\}.$$

In (1zurekstate I) it is shown that

$$G_{\psi} = \theta\,(\Gamma)\,\Upsilon_0 \otimes \Upsilon'_0.$$

The stabilizer in $U(H) \otimes U(K)$ of the ray $\mathbb{C}\,|\psi\rangle$ is $\mathbb{C}_1 G_{\psi}$.

Indeed, if g belongs to this stabilizer, there must be $\lambda \in \mathbb{C}$ such that $g\,|\psi\rangle = \lambda\,|\psi\rangle$. The unitarity of g entails $\lambda \in \mathbb{C}_1$, so that $g \in \mathbb{C}_1 G_{\psi}$.

We now build a matrix representation of G_{ψ}.

We first introduce in each H_{α} an orthonormal basis $|e_{\alpha a}\rangle$, $a \in A_{\alpha}$. The whole set of these kets spans the subspace H. We further introduce an orthonormal basis $|e_b\rangle$, $b \in B$ in $H_{\perp}$, so that the whole set of kets just introduced gives an orthonormal basis for the whole H. Using the isometry η, we can introduce in Im χ the orthonormal basis

$$|\,f_{\alpha a}\rangle = \eta\left(\langle e_{\alpha a}|\right).$$

We then introduce an orthonormal basis $|\,f_c\rangle$, $c \in \mathbb{C}$ in the space Im $\chi_{\perp}$. If $|e\rangle$ is any element of the basis of H and $|\,f\rangle$ is any element of the basis of K, the set of all the elements of the form $|e\rangle \otimes |\,f\rangle$ (shortly $|ef\rangle$) is an orthonormal basis of $U(H) \otimes U(K)$.

Let us express $|\psi\rangle$ by means of this basis. Each χ_{α} sends the $\langle e_{\alpha a}|$ in $|\,f_{\alpha a}\rangle$ and every other element of the basis of H in 0.

Thus $\chi_{\alpha}\left(\langle x|\right) = \sum_{a} \langle x|e_{\alpha a}\rangle\,|\,f_{\alpha a}\rangle$ and then

$$|\chi_{\alpha}\rangle = \sum_{a} |e_{\alpha a}\rangle \otimes |\,f_{\alpha a}\rangle.$$

References

Axiomatics

1. Grangier, P. (2001). Reconstructing the formalism of quantum mechanics in the "contextual objectivity" point of view. arXiv:quant-ph/0111154.
2. Grangier, P. (2002). Contextual objectivity: A realistic interpretation of quantum mechanics. *European Journal of Physics*, *23*(3), 331.
3. Hartle, J. B. (1993). The spacetime approach to quantum mechanics. *Vistas in Astronomy*, *37*, 569–583.
4. Polley, L. (2001). Position eigenstates and the statistical axiom of quantum mechanics. *Foundations of Probability and Physics*, *1*, 314–320.

Bohmian Quantum Mechanics

5. Allori, V., Goldstein, S., Tumulka, R., & Zanghì, N. (2008). On the common structure of Bohmian mechanics and the Ghirardi-Rimini-Weber theory dedicated to GianCarlo Ghirardi on the occasion of his 70th birthday. *The British Journal for the Philosophy of Science*, *59*(3), 353–389.
6. Bohm, D., & Stapp, H. P. (1994). The undivided universe: An ontological interpretation of quantum theory. *American Journal of Physics*, *62*, 958–960.
7. Dürr, D., Goldstein, S., & Zanghì, N. (2004). Quantum equilibrium and the role of operators as observables in quantum theory. *Journal of Statistical Physics*, *116*(1), 959–1055.
8. Mermin, N. D. (2002). Shedding (red and green) light on "time related hidden parameters". arXiv:quant-ph/0206118.

M. Campanella et al., *Interpretative Aspects of Quantum Mechanics*, UNIPA Springer Series,
https://doi.org/10.1007/978-3-030-44207-1

Born's Rule

9. Auffèves, A., & Grangier, P. (2015). A simple derivation of Born's rule with and without Gleason's theorem. arXiv:1505.01369.
10. Barnum, H. (2003). No-signalling-based version of Zurek's derivation of quantum probabilities: A note on "environment-assisted invariance, entanglement, and probabilities in quantum physics". arXiv:quant-ph/0312150.
11. Barnum, H., Caves, C. M., Finkelstein, J., Fuchs, C. A., & Schack, R. (2000). Quantum probability from decision theory? *Proceedings of the Royal Society of London A: Mathematical, Physical and Engineering Sciences*, *456*(1997), 1175–1182 (The Royal Society).
12. Blume-Kohout, R., & Zurek, W. H. (2005). A simple example of "quantum Darwinism": Redundant information storage in many-spin environments. *Foundations of Physics*, *35*(11), 1857–1876.
13. Busch, P. (2003). Quantum states and generalized observables: A simple proof of Gleason's theorem. *Physical Review Letters*, *91*(12), 120403.
14. Campanella, M. (2017). Mathematical studies on quantum mechanics. *Bollettino di Matematica Pura ed applicata*, *9*, 1–49.
15. Gill, R. D. (2003). On an argument of David Deutsch. arXiv:quant-ph/0307188.
16. Hardy, L. (2001). Quantum theory from five reasonable axioms. arXiv:quant-ph/0101012.
17. Hardy, L. (2002). Why quantum theory? *Non-locality and modality* (pp. 61–73). Berlin: Springer.
18. Landsman, N. P. (2009). Born rule and its interpretation. *Compendium of quantum physics* (pp. 64–70). Berlin: Springer.
19. Logiurato, F., & Smerzi, A. (2012). Born rule and noncontextual probability. arXiv:1202.2728.
20. Page, D. N. (2010). Born's rule is insufficient in a large universe. arXiv:1003.2419.
21. Polley, L. (1999). Quantum-mechanical probability from the symmetries of two-state systems. arXiv:quant-ph/9906124.
22. Saunders, S. (2004). Derivation of the Born rule from operational assumptions. *Proceedings of the Royal Society of London A: Mathematical, Physical and Engineering Sciences*, *460*, 1771–1788 (The Royal Society).
23. Schlosshauer, M., & Fine, A. (2005). On Zurek's derivation of the Born rule. *Foundations of Physics*, *35*(2), 197–213.
24. Tipler, F. J. (2006). What about quantum theory? Bayes and the Born interpretation. arXiv:quant-ph/0611245.
25. Wallace, D. (2009). A formal proof of the Born rule from decision-theoretic assumptions. arXiv:0906.2718.
26. Wallach, N. R. (2002). An unentangled Gleason's theorem. *Contemporary Mathematics*, *305*, 291–298.
27. Zurek, W. H. (2005). Probabilities from entanglement, Born's rule $p_k = |\psi_k|^2$ from envariance. *Physical Review A*, *71*(5), 052105.

Categorical Quantum Mechanics

28. Abramsky, S., & Coecke, B. (2008). Categorical quantum mechanics. *Handbook of quantum logic and quantum structures: Quantum logic* (pp. 261–324).
29. Coecke, B., & Perdrix, S. (2010). Environment and classical channels in categorical quantum mechanics. In *International Workshop on Computer Science Logic* (pp. 230–244). Berlin: Springer.
30. Crane, L. (1995). Clock and category: Is quantum gravity algebraic? *Journal of Mathematical Physics*, *36*(11), 6180–6193.
31. Heunen, C., & Vicary, J. (2012). *Lectures on categorical quantum mechanics*. Computer science department. Oxford: Oxford University.

Consistent Histories

32. Gell-Mann, M., & Hartle, J. B. (1994). Equivalent sets of histories and multiple quasiclassical realms. arXiv:gr-qc/9404013.
33. Halliwell, J. J. (1993). Quantum-mechanical histories and the uncertainty principle: Information-theoretic inequalities. *Physical Review D*, *48*(6), 2739.
34. Isham, C. J. (1994). Quantum logic and the histories approach to quantum theory. *Journal of Mathematical Physics*, *35*(5), 2157–2185.
35. Kent, A. (1996). Quasiclassical dynamics in a closed quantum system. *Physical Review A*, *54*(6), 4670.
36. Kent, A. (1996). Remarks on consistent histories and Bohmian mechanics. Bohmian mechanics and quantum theory: An appraisal (pp. 343–352). Berlin: Springer.
37. Kent, A. (1997). Consistent sets yield contrary inferences in quantum theory. *Physical Review Letters*, *78*(15), 2874.
38. Kent, A. (2000). Quantum histories and their implications. *Relativistic quantum measurement and decoherence* (pp. 93–115). Berlin: Springer.

Decision Theory

39. Deutsch, D. (1999). Quantum theory of probability and decisions. *Proceedings of the Royal Society of London A: Mathematical, Physical and Engineering Sciences*, *455*(1998), 3129–3137 (The Royal Society).
40. Wallace, D. (2002). Quantum probability and decision theory, revisited. arXiv:quant-ph/0211104.
41. Wallace, D. (2007). Quantum probability from subjective likelihood: Improving on Deutsch's proof of the probability rule. *Studies in History and Philosophy of Science Part B: Studies in History and Philosophy of Modern Physics*, *38*(2), 311–332.

Decoherence

42. Bacon, D. (2003). Decoherence, control, and symmetry in quantum computers. arXiv:quant-ph/0305025.
43. Blume-Kohout, R., & Zurek, W. H. (2006). Quantum Darwinism: Entanglement, branches, and the emergent classicality of redundantly stored quantum information. *Physical Review A*, *73*(6), 062310.
44. Gell-Mann, M., & Hartle, J. B. (1993). Classical equations for quantum systems. *Physical Review D*, *47*(8), 3345.
45. Gell-Mann, M., & Hartle, J. B. (1995). Strong decoherence. arXiv:gr-qc/9509054.
46. Halliwell, J. J. (1993). Aspects of the decoherent histories approach to quantum mechanics. arXiv:gr-qc/9308005.
47. Halliwell, J. J. (1995). A review of the decoherent histories approach to quantum mechanics. *Annals of the New York Academy of Sciences*, *755*(1), 726–740.
48. Lidar, D. A., & Whaley, K. B. (2003). Decoherence-free subspaces and subsystems. *Irreversible quantum dynamics* (pp. 83–120). Berlin: Springer.
49. Ollivier, H., Poulin, D., & Zurek, W. H. (2005). Environment as a witness: Selective proliferation of information and emergence of objectivity in a quantum universe. *Physical Review A*, *72*(4), 042113.

50. Schlosshauer, M. (2005). Decoherence, the measurement problem, and interpretations of quantum mechanics. *Reviews of Modern Physics*, *76*(4), 1267.
51. Zeh, H. (1997). What is achieved by decoherence? *New developments on fundamental problems in quantum physics* (pp. 441–451). Berlin: Springer.
52. Zurek, W. H. (1986). Reduction of the wavepacket: How long does it take? *Frontiers of nonequilibrium statistical physics* (pp. 145–149). Berlin: Springer.
53. Zurek, W. H. (2003). Decoherence, einselection, and the quantum origins of the classical. *Reviews of Modern Physics*, *75*(3), 715.
54. Zurek, W. H. (2003). Quantum Darwinism and envariance. arXiv:quant-ph/0308163.
55. Zurek, W. H. (2006). Decoherence and the transition from quantum to classical–revisited. *Quantum decoherence* (pp. 1–31). Berlin: Springer.
56. Zurek, W. H. (2007). Relative states and the environment: Einselection, envariance, quantum Darwinism, and the existential interpretation. arXiv:0707.2832.

Entanglement

57. Bell, J. S. (1964). On the Einstein Podolsky Rosen paradox.
58. Bennett, C. H., Bernstein, H. J., Popescu, S., & Schumacher, B. (1996). Concentrating partial entanglement by local operations. *Physical Review A*, *53*(4), 2046.
59. Brukner, C., Zukowski, M., & Zeilinger, A. (2001). The essence of entanglement. arXiv:quant-ph/0106119.
60. Dakic, B., & Brukner, C. (2009). Quantum theory and beyond: Is entanglement special? arXiv:0911.0695.
61. Deutsch, D., & Hayden, P. (2000). Information flow in entangled quantum systems. *Proceedings of the Royal Society of London A: Mathematical, Physical and Engineering Sciences*, *456*(1999), 1759–1774 (The Royal Society).
62. Einstein, A., Podolsky, B., & Rosen, N. (1935). Can quantum-mechanical description of physical reality be considered complete? *Physical Review*, *47*(10), 777.
63. Hill, S., & Wootters, W. K. (1997). Entanglement of a pair of quantum bits. *Physical Review Letters*, *78*(26), 5022.
64. Hooft, G. (2009). Entangled quantum states in a local deterministic theory. arXiv:0908.3408.
65. Horodecki, M., Horodecki, P., & Horodecki, R. (1996). Separability of mixed states: Necessary and sufficient conditions. *Physics Letters A*, *223*(1), 1–8.
66. Horodecki, R., Horodecki, P., Horodecki, M., & Horodecki, K. (2009). Quantum entanglement. *Reviews of Modern Physics*, *81*(2), 865.
67. Korbicz, J., & Lewenstein, M. (2006). Group-theoretical approach to entanglement. *Physical Review A*, *74*(2), 022318.
68. Korbicz, J., & Lewenstein, M. (2007). Remark on a group-theoretical formalism for quantum mechanics and the quantum-to-classical transition. *Foundations of Physics*, *37*(6), 879–896.
69. Korbicz, J., Wehr, J., & Lewenstein, M. (2008). Entanglement of positive definite functions on compact groups. *Communications in Mathematical Physics*, *281*(3), 753–774.
70. Korbicz, J., Wehr, J., & Lewenstein, M. (2009). Entanglement and quantum groups. *Journal of Mathematical Physics*, *50*(6), 062104.
71. Peres, A. (1996). Separability criterion for density matrices. *Physical Review Letters*, *77*(8), 1413.
72. Thompson, C. H. (1996). The chaotic ball: An intuitive analogy for EPR experiments. *Foundations of Physics Letters*, *9*(4), 357–382.
73. Vedral, V., Plenio, M. B., Rippin, M. A., & Knight, P. L. (1997). Quantifying entanglement. *Physical Review Letters*, *78*(12), 2275.

Impact of Quantum Mechanics on the Weltanschauung

74. Brandt, H. E. (2012). Microcausality in quantum field theory. *Physica Scripta, 2012*(T151), 014011.
75. Brun, T. A., Finkelstein, J., & Mermin, N. D. (2002). How much state assignments can differ. *Physical Review A, 65*(3), 032315.
76. De la Peña, L., Cetto, A., & Valdés-Hernández, A. (2012). Quantum behavior derived as an essentially stochastic phenomenon. *Physica Scripta, 2012*(T151), 014008.
77. Duda, J. (2009). Four-dimensional understanding of quantum mechanics. arXiv:0910.2724.
78. Gell-Mann, M., & Hartle, J. B. (1994). Time symmetry and asymmetry in quantum mechanics and quantum cosmology. *Physical Origins of Time Asymmetry, 1*, 311–345.
79. Hooft, G. (1999). Quantum gravity as a dissipative deterministic system. *Classical and Quantum Gravity, 16*(10), 3263.
80. Lokajíček, M., & Kundrát, V. (2012). The controversy between Einstein and Bohr after 75 years, its actual solution and consequences for the present. *Physica Scripta, 2012*(T151), 014007.
81. Mermin, N. (1998). What do these correlations know about reality? Nonlocality and the absurd. arXiv:quant-ph/9807055.
82. Mermin, N. D. (1998). Nonlocal character of quantum theory? *American Journal of Physics, 66*(10), 920–924.
83. Mermin, N. D. (2001). Whose knowledge? arXiv:quant-ph/0107151.
84. Mermin, N. D. (2002). Whose knowledge. *Quantum Theory: Reconsideration of Foundations, 2*, 261–270.
85. Montina, A. (2008). Exponential complexity and ontological theories of quantum mechanics. *Physical Review A, 77*(2), 022104.
86. Norsen, T. (2007). Against 'realism'. *Foundations of Physics, 37*(3), 311–340.
87. Ollivier, H., Poulin, D., & Zurek, W. H. (2004). Objective properties from subjective quantum states: Environment as a witness. *Physical Review Letters, 93*(22), 220401.
88. Pusey, M. F., Barrett, J., & Rudolph, T. (2012). On the reality of the quantum state. *Nature Physics, 8*(6), 475–478.
89. Sudbery, A. (2000). Why am i me? And why is my world so classical? arXiv:quant-ph/0011084.
90. van Dam, W. (2013). Implausible consequences of superstrong nonlocality. *Natural Computing, 12*(1), 9–12.

Hidden Variables

91. Brady, R., & Anderson, R. (2014). Why bouncing droplets are a pretty good model of quantum mechanics. arXiv:1401.4356.
92. Dürr, D., Goldstein, S., & Zanghi, N. (1996). Bohmian mechanics as the foundation of quantum mechanics. *Bohmian mechanics and quantum theory: An appraisal* (pp. 21–44). Berlin: Springer.
93. Tsekov, R. (2009). Bohmian mechanics versus Madelung quantum hydrodynamics. arXiv:0904.0723.

Ithaca Interpretation

94. Angelo, R. M., Brunner, N., Popescu, S., Short, A. J., & Skrzypczyk, P. (2011). Physics within a quantum reference frame. *Journal of Physics A: Mathematical and Theoretical, 44*(14), 145304.
95. Bartlett, S. D., Rudolph, T., & Spekkens, R. W. (2007). Reference frames, superselection rules, and quantum information. *Reviews of Modern Physics, 79*(2), 555.
96. Birman, F. (2009). *Quantum mechanics, correlations, and relational probability (mecánica cuántica, correlaciones y probabilidad relacional)* (pp. 3–22). Crítica: Revista Hispanoamericana de Filosofía.
97. Bitbol, M. (2007). Physical relations or functional relations. *A non-metaphysical construal of Rovelli's relational quantum mechanics*. http://philsci-archive.pitt.edu/3506.
98. Dickson, M. (2004). A view from nowhere: Quantum reference frames and uncertainty. *Studies in History and Philosophy of Science Part B: Studies in History and Philosophy of Modern Physics, 35*(2), 195–220.
99. Esfeld, M. (2003). Do relations require underlying intrinsic properties? A physical argument for a metaphysics of relations. *Metaphysica: International Journal for Ontology and Metaphysics, 4*(1), 5–25.
100. Esfeld, M. (2004). Quantum entanglement and a metaphysics of relations. *Studies in History and Philosophy of Science Part B: Studies in History and Philosophy of Modern Physics, 35*(4), 601–617.
101. Griffiths, R. B. (2003). Probabilities and quantum reality: Are there correlata? *Foundations of Physics, 33*(10), 1423–1459.
102. McCall, S. (2001). The Ithaca interpretation of quantum mechanics, and objective probabilities. *Foundations of Physics Letters, 14*(1), 95–101.
103. Mermin, N. D. (1998). The Ithaca interpretation of quantum mechanics. *Pramana, 51*(5), 549–565.
104. Mermin, N. D. (1998). What is quantum mechanics trying to tell us? *American Journal of Physics, 66*(9), 753–767.
105. Morganti, M. (2009). A new look at relational holism in quantum mechanics. *Philosophy of Science, 76*(5), 1027–1038.
106. Palmer, M. C., Girelli, F., & Bartlett, S. D. (2014). Changing quantum reference frames. *Physical Review A, 89*(5), 052121.
107. Poulin, D., & Yard, J. (2007). Dynamics of a quantum reference frame. *New Journal of Physics, 9*(5), 156.
108. Saunders, S. (1995). Time, quantum mechanics, and decoherence. *Synthese, 102*(2), 235–266.
109. Saunders, S. (1998). Time, quantum mechanics, and probability. *Synthese, 114*(3), 373–404.
110. Seevinck, M. (2006). The quantum world is not built up from correlations. *Foundations of Physics, 36*(10), 1573–1586.
111. Smolin, L. (1995). The Bekenstein bound, topological quantum field theory and pluralistic quantum field theory. arXiv:gr-qc/9508064.
112. Stuckey, W. M., Silberstein, M., & Cifone, M. (2008). Reconciling spacetime and the quantum: Relational blockworld and the quantum liar paradox. *Foundations of Physics, 38*(4), 348–383.
113. Zeilinger, A. (1999). A foundational principle for quantum mechanics. *Foundations of Physics, 29*(4), 631–643.

Levis Principle

114. Ćirković, M. M. (2006). Is quantum suicide painless? On an apparent violation of the principal principle. *Foundations of Science, 11*(3), 287–296.

115. Hawthorne, J., Landes, J., Wallmann, C., & Williamson, J. (2015). The principal principle implies the principle of indifference. *British Journal for the Philosophy of Science, 68*(1), 123–131.
116. Jaeger, L. (2002). Humean supervenience and best-system laws. *International Studies in the Philosophy of Science, 16*(2), 141–155.
117. Karakostas, V. (2009). Humean supervenience in the light of contemporary science. *Metaphysica, 10*(1), 1–26.
118. Lewis, D. (1994). Humean supervenience debugged. *Mind, 103*(412), 473–490.
119. Meacham, C. J. (2005). Three proposals regarding a theory of chance. *Philosophical Perspectives, 19*(1), 281–307.
120. Meacham, C. J. (2009). Two mistakes regarding the principal principle. *The British Journal for the Philosophy of Science, 61*(2), 407–431.
121. Pettigrew, R. (2012). Accuracy, chance, and the principal principle. *Philosophical Review, 121*(2), 241–275.
122. Pigozzi, G. On the notion of admissibility in chance-credence principles: A comment on vranas1.
123. Strevens, M. (1995). A closer look at the 'new' principle. *The British Journal for the Philosophy of Science, 46*(4), 545–561.

Many World Interpretation

124. Baker, D. J. (2007). Measurement outcomes and probability in Everettian quantum mechanics. *Studies in History and Philosophy of Science Part B: Studies in History and Philosophy of Modern Physics, 38*(1), 153–169.
125. Brown, H. R., & Wallace, D. (2005). Solving the measurement problem: De Broglie-Bohm loses out to Everett. *Foundations of Physics, 35*(4), 517–540.
126. Butterfield, J. (2001). Some worlds of quantum theory. arXiv:quant-ph/0105052.
127. Carroll, S. M., & Sebens, C. T. (2014). Many worlds, the Born rule, and self-locating uncertainty. *Quantum theory: A two-time success story* (pp. 157–169). Berlin: Springer.
128. Deutsch, D. (2002). The structure of the multiverse. *Proceedings of the Royal Society of London A: Mathematical, Physical and Engineering Sciences, 458*, 2911–2923 (The Royal Society).
129. DeWitt, B. S. (1971). The many universes interpretation of quantum mechanics. *Foundation of quantum mechanics*. New York: Academic.
130. Donald, M. J. (1997). On many-minds interpretations of quantum theory. arXiv:quant-ph/9703008.
131. Everett, H, III. (1957). "Relative state" formulation of quantum mechanics. *Reviews of Modern Physics, 29*(3), 454.
132. Greaves, H. (2004). Understanding Deutsch's probability in a deterministic multiverse. *Studies in History and Philosophy of Science Part B: Studies in History and Philosophy of Modern Physics, 35*(3), 423–456.
133. Greaves, H. (2007). Probability in the Everett interpretation. *Philosophy Compass, 2*(1), 109–128.
134. Kent, A. (1990). Against many-worlds interpretations. *International Journal of Modern Physics A, 5*(09), 1745–1762.
135. Linde, A., & Vanchurin, V. (2010). How many universes are in the multiverse? *Physical Review D, 81*(8), 083525.
136. Osnaghi, S., Freitas, F., & Freire, O, Jr. (2009). The origin of the Everettian heresy. *Studies in History and Philosophy of Science Part B: Studies in History and Philosophy of Modern Physics, 40*(2), 97–123.

137. Page, D. N. (1999). Can quantum cosmology give observational consequences of many-worlds quantum theory? *AIP Conference Proceedings, AIP, 493*, 225–232.
138. Rubin, M. A. (2004). There is no basis ambiguity in Everett quantum mechanics. *Foundations of Physics Letters, 17*(4), 323–341.
139. Saunders, S., & Wallace, D. (2008). Branching and uncertainty. *The British Journal for the Philosophy of Science, 59*(3), 293–305.
140. Sebens, C. T., & Carroll, S. M. (2016). Self-locating uncertainty and the origin of probability in Everettian quantum mechanics. *The British Journal for the Philosophy of Science, 69*(1), 25–74.
141. Stapp, H. P. (2002). The basis problem in many-worlds theories. *Canadian Journal of Physics, 80*(9), 1043–1052.
142. Tegmark, M. (1998). The interpretation of quantum mechanics: Many worlds or many words? *Fortschritte der Physik: Progress of Physics, 46*(6–8), 855–862.
143. Wallace, D. (2002). Worlds in the Everett interpretation. *Studies in History and Philosophy of Science Part B: Studies in History and Philosophy of Modern Physics, 33*(4), 637–661.
144. Wallace, D. (2003). Everett and structure. *Studies in History and Philosophy of Science Part B: Studies in History and Philosophy of Modern Physics, 34*(1), 87–105.
145. Wallace, D. (2003). Everettian rationality: Defending Deutsch's approach to probability in the Everett interpretation. *Studies in History and Philosophy of Science Part B: Studies in History and Philosophy of Modern Physics, 34*(3), 415–439.
146. Wallace, D. (2006). Epistemology quantized: Circumstances in which we should come to believe in the Everett interpretation. *The British Journal for the Philosophy of Science, 57*(4), 655–689.
147. Zurek, W. H. (2009). Quantum Darwinism. *Nature Physics, 5*(3), 181–188.

Measurements

148. Allahverdyan, A. E., Balian, R., & Nieuwenhuizen, T. M. (2013). Understanding quantum measurement from the solution of dynamical models. *Physics Reports, 525*(1), 1–166.
149. Franke, K., Bednorz, A., & Belzig, W. (2012). Time asymmetry in weak measurements. *Physica Scripta, 2012*(T151), 014013.
150. Mermin, N. D. (1999). A Kochen-Specker theorem for imprecisely specified measurement. arXiv:quant-ph/9912081.
151. Poulin, D. (2005). Macroscopic observables. *Physical Review A, 71*(2), 022102.

Modal Interpretation

152. de Ronde, C. (2010). For and against metaphysics in the modal interpretation of quantum mechanics. *Philosophica, 83*, 85–117.
153. Dieks, D. (1995). Physical motivation of the modal interpretation of quantum mechanics. *Physics Letters A, 197*(5–6), 367–371.
154. Dieks, D. (2007). Probability in modal interpretations of quantum mechanics. *Studies in History and Philosophy of Science Part B: Studies in History and Philosophy of Modern Physics, 38*(2), 292–310.
155. Lombardi, O., & Castagnino, M. (2008). A modal-Hamiltonian interpretation of quantum mechanics. *Studies in History and Philosophy of Science Part B: Studies in History and Philosophy of Modern Physics, 39*(2), 380–443.

Quantum Mechanics and Information Theory

156. Acín, A., Augusiak, R., Cavalcanti, D., Hadley, C., Korbicz, J. K., Lewenstein, M., et al. (2010). Unified framework for correlations in terms of local quantum observables. *Physical Review Letters, 104*(14), 140404.
157. Barnum, H., Beigi, S., Boixo, S., Elliott, M. B., & Wehner, S. (2010). Local quantum measurement and no-signaling imply quantum correlations. *Physical Review Letters, 104*(14), 140401.
158. Belavkin, V. P., & Ohya, M. (2002). Entanglement, quantum entropy and mutual information. *Proceedings of the Royal Society of London A: Mathematical, Physical and Engineering Sciences, 458*, 209–231 (The Royal Society).
159. Deutsch, D. E., Barenco, A., & Ekert, A. (1995). Universality in quantum computation. *Proceedings of the Royal Society of London A, 449*(1937), 669–677.
160. Devine, S. (2009). The insights of algorithmic entropy. *Entropy, 11*(1), 85–110.
161. DiVincenzo, D. P. (1998). Quantum gates and circuits. *Proceedings of the Royal Society of London A: Mathematical, Physical and Engineering Sciences, 454*, 261–276 (The Royal Society).
162. Fuchs, C. A. (2002). Quantum mechanics as quantum information (and only a little more). arXiv:quant-ph/0205039.
163. Fuchs, C. A., et al. (2001). Quantum foundations in the light of quantum information. *NATO Science Series Sub Series III Computer and Systems Sciences, 182*, 38–82.
164. Kowalski, A. M., Plastino, A., & Casas, M. (2009). Generalized complexity and classical-quantum transition. *Entropy, 11*(1), 111–123.
165. Mermin, N. D. (2001). From classical state swapping to quantum teleportation. *Physical Review A, 65*(1), 012320.
166. Rains, E. M., Hardin, R., Shor, P. W., & Sloane, N. (1997). A nonadditive quantum code. *Physical Review Letters, 79*(5), 953.
167. Timpson, C. G. (2003). On a supposed conceptual inadequacy of the Shannon information in quantum mechanics. *Studies in History and Philosophy of Science Part B: Studies in History and Philosophy of Modern Physics, 34*(3), 441–468.
168. Vidal, G., & Cirac, J. I. (2000). Storage of quantum dynamics on quantum states: A quasi-perfect programmable quantum gate. arXiv:quant-ph/0012067.

Quantum Mechanics, Quantum Logic and Probability

169. Barnum, H., Fuchs, C. A., Renes, J. M., & Wilce, A. (2005). Influence-free states on compound quantum systems. arXiv:quant-ph/0507108.
170. Belavkin, V. P. (1985). Reconstruction theorem for a quantum stochastic process. *Theoretical and Mathematical Physics, 62*(3), 275–289.
171. Belavkin, V. P. (2005). Quantum diffusion, measurement and filtering. arXiv:quant-ph/0510028.
172. Caves, C. M., & Schack, R. (2005). Properties of the frequency operator do not imply the quantum probability postulate. *Annals of Physics, 315*(1), 123–146.
173. Hess, K., De Raedt, H., & Michielsen, K. (2012). Hidden assumptions in the derivation of the theorem of Bell. *Physica Scripta, 2012*(T151), 014002.
174. Khrennikov, A. (2012). Violation of Bell's inequality by correlations of classical random signals. *Physica Scripta, 2012*(T151), 014003.
175. Pitowsky, I. (2006). Quantum mechanics as a theory of probability. *Physical theory and its interpretation* (pp. 213–240). Berlin: Springer.
176. Popescu, S. (1995). Bell's inequalities and density matrices: Revealing 'hidden' nonlocality. *Physical Review Letters, 74*(14), 2619.

177. Rédei, M., & Summers, S. J. (2007). Quantum probability theory. *Studies in History and Philosophy of Science Part B: Studies in History and Philosophy of Modern Physics, 38*(2), 390–417.
178. Richman, F., & Bridges, D. (1999). A constructive proof of Gleason's theorem. *Journal of Functional Analysis, 162*(2), 287–312.
179. Saunders, S. (2005). What is probability? *Quo Vadis quantum mechanics?* (pp. 209–238). Berlin: Springer.

Relational Quantum Mechanics

180. Griffiths, R. B. (1996). Consistent histories and quantum reasoning. *Physical Review A, 54*(4), 2759.
181. Laudisa, F. (2001). The EPR argument in a relational interpretation of quantum mechanics. *Foundations of Physics Letters, 14*(2), 119–132.
182. Rovelli, C. (1996). Relational quantum mechanics. *International Journal of Theoretical Physics, 35*(8), 1637–1678.
183. Smerlak, M., & Rovelli, C. (2007). Relational EPR. *Foundations of Physics, 37*(3), 427–445.

Additional Bibliography

184. Allahverdyan, A. E., Balian, R., & Nieuwenhuizen, T. M. (2011). Understanding quantum measurement from the solution of dynamical models. arXiv:1107.2138.
185. Assis, A. V. (2011). On the nature of a* kak and the emergence of the Born rule. *Annalen der Physik, 523*(11), 883–897.
186. Barrett, J. A. (2009). Ithaca interpretation of quantum mechanics. *Compendium of Quantum Physics* (pp. 325–326). Berlin: Springer.
187. Berndl, K., Daumer, M., Dürr, D., Goldstein, S., & Zanghì, N. (1995). A survey of Bohmian mechanics. *Il Nuovo Cimento B (1971-1996), 110*(5–6), 737–750.
188. Blanchard, P., & Brüning, E. (2006). Remarks on the structure of states of composite quantum systems and envariance. *Physics Letters A, 355*(3), 180–187.
189. Blaylock, G. (2010). The EPR paradox, Bell's inequality, and the question of locality. *American Journal of Physics, 78*(1), 111–120.
190. Blood, C. (2013). Derivation of the coefficient squared probability law in quantum mechanics. arXiv:1306.0213.
191. Bolotin, A. (2016). Explaining the Born rule in the intuitionistic interpretation of quantum mechanics. arXiv:1610.01847.
192. Brumer, P., & Gong, J. (2006). Born rule in quantum and classical mechanics. *Physical Review A, 73*(5), 052109.
193. Bub, J. (1999). *Interpreting the quantum world.* Cambridge: Cambridge University Press.
194. Campbell, J. O. (2016). Universal Darwinism as a process of Bayesian inference. *Frontiers in Systems Neuroscience, 10.*
195. Deffner, S. (2016). Demonstration of entanglement assisted invariance on IBM's quantum experience. arXiv:1609.07459.
196. Deffner, S., & Zurek, W. H. (2016). Foundations of statistical mechanics from symmetries of entanglement. *New Journal of Physics, 18*(6), 063013.
197. DeWitt, B. S. (1971). The many-universes interpretation of quantum mechanics. In B. d'Espagnat (Ed.), *Foundations of quantum mechanics*. New York: Academic.

198. Dikshit, B. (2017). A simple proof of Born's rule for statistical interpretation of quantum mechanics. *Journal for Foundations and Applications of Physics*, *4*(1), 24–30.
199. Dressel, J. (2013). *Indirect observable measurement: An algebraic approach*. Ph.D. thesis, University of Rochester.
200. Feng, Y.-L., & Chen, Y.-X. (2014). "Almost" quotient space, non-dynamical decoherence and quantum measurement. arXiv:1409.7167.
201. Fields, C. (2011). Quantum mechanics from five physical assumptions. arXiv:1102.0740.
202. Fields, C. (2013). Imposing system-observer symmetry on the von Neumann description of measurement. arXiv:1308.1383.
203. Fields, C. (2014). Consistent quantum mechanics admits no mereotopology. *Axiomathes*, *24*(1), 9–18.
204. Fields, C. (2014). On the Ollivier-Poulin-Zurek definition of objectivity. *Axiomathes*, *24*(1), 137–156.
205. Frigg, R., & Werndl, C. (2011). A guide for the perplexed. *Probabilities in physics* (p. 115).
206. Galley, T. D., & Masanes, L. (2016). Classification of all alternatives to the Born rule in terms of informational properties. arXiv:1610.04859.
207. Gleason, A. M. (1957). Measures on the closed subspaces of a Hilbert space. *Journal of Mathematics and Mechanics*, *6*(6), 885–893.
208. Harris, J., Bouchard, F., Santamato, E., Zurek, W. H., Boyd, R. W., & Karimi, E. (2016). Quantum probabilities from quantum entanglement: Experimentally unpacking the Born rule. *New Journal of Physics*, *18*(5), 053013.
209. Hasse, C. L. (2014). *On the individuation ofphysical systems in quantum theory*. Ph.D. thesis.
210. Herbut, F. (2007). Derivation of the quantum probability law from minimal non-demolition measurement. *Journal of Physics A: Mathematical and Theoretical*, *40*(34), 10549.
211. Herbut, F. (2007). Quantum probability law from 'environment-assisted invariance' in terms of pure-state twin unitaries. *Journal of Physics A: Mathematical and Theoretical*, *40*(22), 5949.
212. Herbut, F. (2012). Zurek's envariance derivation of Born's rule and measurement. *The European Physical Journal Plus*, *127*(2), 1–7.
213. Joos, E., Zeh, H. D., Kiefer, C., Giulini, D. J., Kupsch, J., & Stamatescu, I.-O. (2013). *Decoherence and the appearance of a classical world in quantum theory*. New York: Springer Science & Business Media.
214. Jordan, T. F. (2006). Assumptions that imply quantum dynamics is linear. *Physical Review A*, *73*(2), 022101.
215. Jordan, T. F. (2009). Why quantum dynamics is linear. *Journal of Physics: Conference Series*, *196*(1), 012010 (IOP Publishing).
216. Kent, A. (2010). One world versus many: The inadequacy of Everettian accounts of evolution, probability, and scientific confirmation. *Many worlds* (pp. 307–354).
217. Landsman, N. P. (2009). Algebraic quantum mechanics. *Compendium of quantum physics* (pp. 6–10). Berlin: Springer.
218. Lavis, D. (2011). An objectivist account of probabilities in statistical mechanics. In C. Beisbart & S. Hartmann (Eds.), *Probabilities in physics* (pp. 51–81). Oxford: Oxford University Press.
219. Leifer, M. (2015). "It from bit" and the quantum probability rule. *It from bit or bit from it?* (pp. 5–23). Berlin: Springer.
220. Lesovik, G. (2014). Derivation of the Born rule from the unitarity of quantum evolution. arXiv:1411.6992.
221. Mohrhoff, U. (2004). Probabilities from envariance? *International Journal of Quantum Information*, *2*(02), 221–229.
222. Moldoveanu, F. (2014). Quantum mechanics reconstruction from invariance of the laws of nature under tensor composition. arXiv:1407.7610.
223. Moldoveanu, F. (2016). Unitary realization of wave function collapse. *International Journal of Quantum Information*, *14*(04), 1640015.
224. Nenashev, A. (2014). Quantum probabilities from combination of Zurek's envariance and Gleason's theorem. *Physica Scripta*, *2014*(T163), 014033.

225. Nenashev, A. (2016). Why state of quantum system is fully defined by density matrix. arXiv:1601.08205.
226. Nenashev, A. V. (2013). Derivation of the Born rule based on the minimal set of assumptions. arXiv:1308.5384.
227. Olivares, S., & Paris, M. G. (2009). Entanglement-induced invariance in bilinear interactions. *Physical Review A*, *80*(3), 032329.
228. Randall, A. F. (2016). Quantum probability as an application of data compression principles. arXiv:1606.06802.
229. Rashkovskiy, S. A. (2016). Quantum mechanics without quanta: The nature of the wave-particle duality of light. *Quantum Studies: Mathematics and Foundations*, *3*(2), 147–160.
230. Riedel, C. J., Zurek, W. H., & Zwolak, M. (2016). Objective past of a quantum universe: Redundant records of consistent histories. *Physical Review A*, *93*(3), 032126.
231. Rubin, M. A. (2001). Locality in the Everett interpretation of Heisenberg-picture quantum mechanics. *Foundations of Physics Letters*, *14*(4), 301–322.
232. Scheibe, E., & Sykes, J. (1973). *The logical analysis of quantum mechanics*. Oxford: Pergamon Press.
233. Schlosshauer, M. (2006). Experimental motivation and empirical consistency in minimal no-collapse quantum mechanics. *Annals of Physics*, *321*(1), 112–149.
234. Schlosshauer, M. A. (2007). *Decoherence: And the quantum-to-classical transition*. New York: Springer Science & Business Media.
235. Seidewitz, E. (2011). Consistent histories of systems and measurements in spacetime. *Foundations of Physics*, *41*(7), 1163–1192.
236. Son, W. (2014). Consistent theory for causal non-locality beyond the Born's rule. *Journal of the Korean Physical Society*, *64*(4), 499–503.
237. Stoica, O. C. (2016). Quantum measurement and initial conditions. *International Journal of Theoretical Physics*, *55*(3), 1897–1911.
238. Streltsov, A., & Zurek, W. H. (2013). Quantum discord cannot be shared. *Physical Review Letters*, *111*(4), 040401.
239. Tanona, S. (2013). Decoherence and the Copenhagen cut. *Synthese*, 1–25.
240. Tell, A. O. (2012). A theory of quantum observation and the emergence of the Born rule. arXiv:1205.0293.
241. van Wezel, J. (2008). Quantum dynamics in the thermodynamic limit. *Physical Review B*, *78*(5), 054301.
242. Vermeyden, L., Ma, X., Lavoie, J., Bonsma, M., Sinha, U., Laflamme, R., et al. (2014). An experimental test of envariance. arXiv:1408.7087.
243. Vermeyden, L., Ma, X., Lavoie, J., Bonsma, M., Sinha, U., Laflamme, R., et al. (2015). Experimental test of environment-assisted invariance. *Physical Review A*, *91*(1), 012120.
244. Wallace, D. Language use in a branching universe.
245. Weimer, H. A quantum-thermodynamic approach to transport phenomena.
246. Wüthrich, C. (2011). Can the world be shown to be indeterministic after all? *Probabilities in physics* (pp. 365–390).
247. Zurek, W. (2004). Quantum Darwinism and envariance science and ultimate reality: From quantum to cosmos. In J. D. Barrow, P. C. W. Davies, & Ch. Harper.
248. Zurek, W. H. (1981). Pointer basis of quantum apparatus: Into what mixture does the wave packet collapse? *Physical Review D*, *24*(6), 1516.
249. Zurek, W. H. (1992). The environment, decoherence, and the transition from quantum to classical. In *Quantum Gravity and Cosmology-Proceedings of the Xxii Gift International Seminar on Theoretical Physics* (p. 117). World Scientific.
250. Zurek, W. H. (1998). Decoherence, einselection and the existential interpretation (the rough guide). *Proceedings of the Royal Society of London A: Mathematical, Physical and Engineering Sciences*, 1793–1822.
251. Zurek, W. H. (2003). Environment-assisted invariance, entanglement, and probabilities in quantum physics. *Physical Review Letters*, *90*(12), 120404.

252. Zurek, W. H. (2007). Quantum origin of quantum jumps: Breaking of unitary symmetry induced by information transfer in the transition from quantum to classical. *Physical Review A*, *76*(5), 052110.
253. Zurek, W. H. (2011). Entanglement symmetry, amplitudes, and probabilities: Inverting Born's rule. *Physical Review Letters*, *106*(25), 250402.
254. Zurek, W. H. (2013). Quantum theory of the classical: Einselection, envariance, and quantum Darwinism. Los Alamos National Laboratory Doc no: LA-UR-13-22611.
255. Zurek, W. H. (2013). Wave-packet collapse and the core quantum postulates: Discreteness of quantum jumps from unitarity, repeatability, and actionable information. *Physical Review A*, *87*(5), 052111.
256. Zurek, W. H. (2014). Quantum Darwinism, decoherence, and the randomness of quantum jumps. *Physics Today*, *67*, 44–50.

Printed in the United States
by Baker & Taylor Publisher Services

Printed in the United States
by Baker & Taylor Publisher Services